Nanostructured Materials for Electromagnetic Interference Shielding

Nanostructured Materials for Electromagnetic Interference Shielding

Edited by
Sabu Thomas
Suji Mary Zachariah

CRC Press is an imprint of the
Taylor & Francis Group, an **informa** business

First edition published 2022
by CRC Press
6000 Broken Sound Parkway NW, Suite 300, Boca Raton, FL 33487–2742

and by CRC Press
2 Park Square, Milton Park, Abingdon, Oxon, OX14 4RN

CRC Press is an imprint of Taylor & Francis Group, LLC

ISBN: 978-1-032-10836-0 (hbk)
ISBN: 978-1-032-10842-1 (pbk)
ISBN: 978-1-003-21731-2 (ebk)

DOI: 10.1201/9781003217312

Typeset in Times LT Std
by Apex CoVantage, LLC

CONTENTS

Preface vii

Editors' Biographies ix

List of Contributors xi

1 An Overview of Electromagnetic Interference Shielding 1
Aastha Dutta

2 Carbon Materials: Potential Agents in Electromagnetic Interference Shielding 9
Suji Mary Zachariah, P. A. Nizam, Yves Grohens, Nandakumar Kalarikkal, and Sabu Thomas

3 Novel MXene Materials for Electromagnetic Interference Shielding Applications 23
P. A. Nizam, Suji Mary Zachariah, and Sabu Thomas

4 Biodegradable Thermoplastic Polymer Nanocomposites for Screening Electromagnetic Radiation 35
Gopika G. Nair and Preema C. Thomas

5 Conducting Polymer-Based Materials for Electromagnetic Interference Shielding Applications 55
Deepa K. Baby

6 Absorption- and Reflection-Dominated Materials for Electromagnetic Interference Shielding Applications 65
Lavanya Jothi

7 High-Temperature Materials for Electromagnetic Interference Shielding Applications 77
Ragin Ramdas M.

8 Electromagnetic Interference Shielding Materials for Aerospace Applications 87
Ananthu Prasad, Miran Mozetič, and Sabu Thomas

9 Overview of Nanostructured Materials for Electromagnetic Interference Shielding 99
P. A. Nizam, T. Binumol, and Sabu Thomas

Index 111

PREFACE

There has been tremendous growth in the usage and application of electronic devices across the globe in many sectors, including industrial, commercial, and military, which has led to the creation of a new form of pollution known as electromagnetic interference (EMI) or electromagnetic smog. Its effects range from minor disturbances in broadcast reception to potentially fatal accidents due to the corruption of safety-critical control systems. It may also cause electrical and electronic malfunctioning, can ignite flammable or other hazardous atmospheres, and can have a direct effect on human health. As electronic systems penetrate more deeply into all aspects of society, both the potential for interference effects and the potential for serious EMI-induced incidents will increase. Therefore, this has become a serious problem and its mitigation could be achieved through the use of EMI shielding materials, which prevent the transmission of EM radiation by reflection and/or absorption of the electromagnetic radiation or by suppressing the radiation. This book is motivated by the fact that there is an increase in attention to electromagnetic pollution issues arising from the rapidly growing need for and usage of electronic and electrical devices. With this in mind, this book is designed to give a comprehensive overview of the sources and effects of EM radiation and various novel materials employed for preventing it. Therefore, this book is important because it explores all aspects of electromagnetic shielding materials and examines the current state of the art and new challenges in this rapidly growing area.

Therefore, it is significant and timely to describe various nanomaterials for EMI shielding. The authors discuss in detail the fundamentals of shielding theory, the practice of electromagnetic field measuring techniques, some of the EMI standards, novel materials employed (like MXenes), and the application of these materials in various fields.

The editors have made a conscious effort to select authors representing diverse disciplines, and we would like to thank them profusely for their high-quality submissions and for contributing to this truly multidisciplinary effort. We offer special thanks to our readers and to the editorial staff of the Taylor & Francis Group/CRC Press for their assistance and helpful suggestions at every step.

EDITORS' BIOGRAPHIES

Suji Mary Zachariah
Miss Suji Mary Zachariah is currently working as a research scholar under the guidance of Professor Sabu Thomas at Mahatma Gandhi University, Kottayam, Kerala, India. She has expertise in polymer science, and her research interest lies in the field of polymer-based nanocomposites for electromagnetic interference shielding application. She mainly works toward the development of absorption-dominated materials from sustainable resources and the fabrication of several biopolymer composites with improved mechanical, electrical, and shielding properties.

Professor Dr. Sabu Thomas
Professor Sabu Thomas is currently Vice-Chancellor of Mahatma Gandhi University and Founder Director and Professor of the International and Inter University Centre for Nanoscience and Nanotechnology. He has been a full professor of Polymer Science and Technology in the School of Chemical Sciences, Mahatma Gandhi University, Kottayam, Kerala, India. Prof. Thomas is an outstanding leader who has received sustained international acclaim for his work in nanoscience, polymer science and engineering, polymer nanocomposites, elastomers, polymer blends, interpenetrating polymer networks, polymer membranes, green composites and nanocomposites, nanomedicine, and green nanotechnology. Prof. Thomas has received a Honoris Causa DSc from the University of South Brittany, France, and the University of Lorraine, France. Very recently, he was named a Foreign Fellow of the European Academy of Sciences (EurASc) and was listed in the top 2% of scientists in

India by Stanford University. He has received many national and international awards and has published over 1,000 peer-reviewed research papers, reviews, and book chapters. He has co-edited 80 books and is the holder of more than five patents (four granted, eleven filed).

CONTRIBUTORS

Deepa K. Baby
Department of Basic Sciences and Humanities
Rajagiri School of Engineering and Technology (Autonomous)
Kakkanad, Kerala, India
deepakb@rajagiritech.edu.in

T. Binumol
School of Chemical Sciences
Mahatma Gandhi University
Kottayam, Kerala, India
binumoltanunivas999@gmail.com

Aastha Dutta
Maharashtra Institute of Technology
Aurangabad, India
aastha2006@gmail.com

Yves Grohens
Centre de Recherche Christian Huygens
Rue de Saint-Maudé, Lorient, France
yves.grohens@univ-ubs.fr

Lavanya Jothi
Department of Physics, Saveetha School of Engineering
Saveetha Institute of Medical and Technical Sciences
Saveetha University
Chennai, Tamilnadu, India
lavanyajth@gmail.com

Nandakumar Kalarikkal
School of Pure and Applied Physics and International and Inter University Centre for Nanoscience and Nanotechnology (IIUCNN)
Mahatma Gandhi University
Kottayam, Kerala, India
nkkalarikkal@mgu.ac.in

Miran Mozetič
Jozef Stefan Institute
Ljubljana, Slovenia
miran.mozetic@guest.arnes.si

Gopika G. Nair
Department of Physics
CMS College (Autonomous)
Kottayam, Kerala, India
gopikagnair2611@gmail.com

P. A. Nizam
School of Chemical Sciences
Mahatma Gandhi University
Kottayam, Kerala, India
nizam000@gmail.com

Ananthu Prasad
International and Inter University Centre for Nanoscience and Nanotechnology (IIUCNN)
Mahatma Gandhi University
Kottayam, Kerala, India
ananthuprasad069@gmail.com

Ragin Ramdas M.
Department of Basic Sciences and Humanities
Rajagiri School of Engineering and Technology (Autonomous)
Kakkanad, Kerala, India
ragram53@gmail.com

Preema C. Thomas
Department of Physics
CMS College (Autonomous)
Kottayam, Kerala, India
preema@cmscollege.ac.in

Sabu Thomas
International and Inter University Centre for Nanoscience and Nanotechnology (IIUCNN)
Mahatma Gandhi University
Kottayam, Kerala, India
sabuthomas@mgu.ac.in

Suji Mary Zachariah
International and Inter University Centre for Nanoscience and Nanotechnology (IIUCNN)
Mahatma Gandhi University
Kottayam, Kerala, India
Centre de Recherche Christian Huygens
Rue de Saint-Maudé, Lorient, France
sujimaryzachariah@gmail.com

1

AN OVERVIEW OF ELECTROMAGNETIC INTERFERENCE SHIELDING

Aastha Dutta

⇒ **CONTENTS**

1.1 Introduction 1
1.2 Theory of EMI Shielding 2
1.3 Materials for EMI Shielding 4
1.4 Conclusion 7
References 7

1.1 INTRODUCTION

In today's progressive world, the demand for electronic devices is increasing by leaps and bounds. But as their usage is increasing, recognition of some of their drawbacks is also simultaneously increasing. One major drawback that has been recognized for these devices is electromagnetic interference. EMI is the result of the higher packing density of these devices [1], [2]. On the one hand, this packing pattern enables quick response by the device, but on the other hand, the magnetic interference can result in the device malfunctioning [3]. Also, the interference results in damage to the communication system and safety parameters of many devices. The most relevant cause of EMI is electrostatic discharge (ESD). It can be realized in the form of flashes on a television screen or a clicking sound on audio systems. There are some adverse effects of EMI on human health, such as insomnia, nervousness, and headache, upon being exposed to these electromagnetic waves [4], [5], [6]. In the operation of microprocessor controller devices, high-frequency signals are emitted in the surrounding environment and result in damage to or malfunction of the surrounding equipment. In order to prevent the previously mentioned

DOI: 10.1201/9781003217312-1

damage or hazards, these types of interference must be managed and filtered by means of EMI shielding. In this chapter, an overview of the various methods and materials being used for EMI shielding will be discussed.

1.2 THEORY OF EMI SHIELDING

Electromagnetic waves have long wavelengths and frequencies in the GHz range. The shielding materials are basically chosen on the basis of these long wavelengths [7]. The ratio of impinging energy to the residual energy is measured in terms of shielding effectiveness. When an electromagnetic wave passes through a shield, part of it is absorbed, part of it is reflected, and the unaffected waves emerge out from the shield as residual energy. The reflected radiation is undesirable, and for safety-related reasons the most desirable phenomenon is absorption of the rays. Electromagnetic waves are studied by means of two parameters: magnetic field, denoted by "*H*," and electric field, denoted by "*E*." *E* and *H* are perpendicular to each other, and the waves propagate at right angles to the plane containing these two fields. The ratio of electric field to magnetic field is known as wave impedance [8]. The reflection of waves can be reduced by decreasing this value. But normally, the shielding materials are made up of conducting materials such as metals and carbon; hence, they show variation in the values. By adding nanomaterials to these shielding materials, the impedance issue is sorted out to some extent. The EMI shielding region is categorized into two ranges, namely, the near field shielding region and the far [9] field shielding region. In the far field shielding region, the distance between the radiation source and the EMI shield is greater than $\lambda/2\pi$ and vice versa [10].

The change in power that occurs due to the interaction of incident radiation and the shielding material is referred to as power loss. It consists of both magnetic and electric field losses. Power loss depends upon the electrical and magnetic properties of the materials. It also depends upon the respective amplitudes of the electric and magnetic fields in the incident radiation. This power loss is also known as shielding effectiveness or total shielding effectiveness (SE_T), and it is measured in decibels. The power loss due to absorption is called shielding effectiveness absorption loss (SE_A); similarly, for refection-related loss it is called shielding effectiveness reflection loss (SE_R) [7]. The loss that stems from multiple reflections from the interior of the sample is denoted by $SE_{M.}$ The overall effectiveness of shielding, SE_T, is calculated by using the following formula:

$$SE_T = SE_A + SE_R + SE_M \quad (1.1)$$
$$SE_T = -10 \log (T) \quad (1.2)$$
$$SE_R = -10 \log (1 - R) \quad (1.3)$$

If the value of total shielding effectiveness exceeds 15 decibels, then the loss from multiple reflections is taken to be negligible. Equation 1.1 then becomes

$$SE_A = SE_T - SE_R = -10 \log (T/(1 - R)) \quad (1.4)$$

At a given frequency, the absorption loss increases with both permeability and conductivity. The magnetic property of the material enhances its radiation absorption capacity because of the strong interaction of its magnetic dipoles with the radiation's magnetic field [7].

The major factors considered in magnetic domains are dipole-dipole interactions, magnetic susceptibility and continuity, and magnetic dipole friction. Similarly, the electrical property of the material affects its absorption capacity. Here, the important properties to be considered are dipole-dipole interaction, dielectric connectivity, electric susceptibility, and electric dipole fraction. The conducting property of a material, for example, carbon, is directly related to its dielectric behavior. Therefore, when carbon-containing materials are treated at different temperatures, their conductivity and permittivity increase with increasing temperature. If a shielding material is nonmagnetic in nature, then its high conductivity is utilized for obtaining high shielding effects.

The surface area of the shielding material is known as skin effect, and it helps to enhance the shielding effect by increasing the interaction of the volume of the part of the specimen with the radiation. The area lying between the shielding component (filler) and the nonshielding component is termed the interfacial area. The nonshielding material (matrix) shows a low degree of interaction with the radiation. On other hand, if the material is porous, then it has a higher surface area and shows a higher degree of shielding. The particle size of the filler used is inversely proportional to the shielding power of that material. If the shielding material, that is, the filler, is magnetic in nature, then it should show magnetic continuity so that its magnetic field lines are continuous. So, to achieve a high degree of shielding, magnetic percolation of the composite is required. All of the properties discussed so far are dependent on the frequency of the radiation and the temperature. The absorption mechanism shows drastic changes for some materials as the frequency changes. When materials were compared, it was observed that because ceramics have slower ionic movements, their frequency-dependent behavior is more prominent than that of carbon or metals. When reflection of radiation is considered, low reflection materials are preferred for safety reasons. For this reason, magnetic metallic materials that have high conductivity as well as high impedance are not preferred as they show high reflectance also. In this case, to reduce the amount of reflection, ceramic magnetic materials are being used [7].

Another factor that promotes shielding is the electrical connectivity nature of the material. This electrical continuity also helps with the flow of the eddy currents that are induced because of the magnetic field. This subsequently promotes the loss of magnetic power. Therefore, if a composite is prepared using a conductive filler and a nonconductive matrix, then it shows good shielding properties. As the aspect ratio of the filler increases (particle size decreases), composite materials show lower percolation threshold values. When electromagnetic radiation is absorbed by a body, the rise in the temperature of that body is due to the conversion of this energy to thermal energy. This temperature change affects the extent of the absorption and reflection of that body. In certain materials, for example, graphene, it was observed that the temperature rise enhanced not only its absorption capacity but also its shielding capacity [11].

1.3 MATERIALS FOR EMI SHIELDING

For EMI shielding, the main properties needed are high conductivity and mobile electrons for interacting with radiation. The most preferred materials for this application are carbons, ceramics, metals, conducting polymers, cement, or concrete. Normally, an EMI shielding material is composed of a filler and a matrix and falls in the category of composites. As cement and ceramics only have ions for interacting with radiation, they prove to be less preferable for this use. Similarly, polymers can be chosen only if they have high conductivity. As discussed previously, the absorption of any material, and also its shielding capacity, is enhanced by the presence of magnetic constituents in it. In terms of cost-effectiveness, the composites have turned out to be the most preferable materials. If the material being used is something other than a composite, then its mode of manufacture is tedious and expensive. Flexible graphite is an example of such a material.

The shielding materials are classified into two categories, namely, structured and functional materials. Functional materials are used in function-specific applications, such as in cell phones. The structured materials have good-load bearing capacity. For example, in airframes, continuous carbon fiber polymer-matrix composites are being used for their light weight. Another example is the cement-based, concrete cover of a large transformer vault. Some structured materials have multifunctional capacities. Compared to functional materials, structured materials have the following advantages: they are economical, durable, have a high functioning capacity, and do not show any deterioration in mechanical properties with time. When functional materials are used, then their density and thickness need to be monitored. But for applications such as aircraft, lower density materials cannot provide effective shielding. Also, in civil infrastructures, low-density materials cannot satisfy the requirements. Hence, the usage of functional materials is hindered because all of their properties do not satisfy the end user requirements.

Metals being used for shielding applications are either coated or used in their bulk form. Aluminum sheets used for enclosing electronic goods is an example of its bulk form. Nickel deposition by electroplating uses metal in the coated form. Both of these forms have disadvantages, namely, the bulk form has large mass and volume and its enclosure suffers from electromagnetic shielding deficiency at the seams. On the other hand, the coated form becomes scratched when exposed to radiation. The magnetic metals have high absorption as well as high shielding capacity. Examples are mu-metal, permalloy, and stainless steel. All of these alloys have the disadvantage of high density. A polymer matrix composite, for example, iron-nickel particles coated with carbon nanotubes, is very effective in this field [12]. The polymer matrix composite shows higher shielding capacity compared to the composite that has the magnetic component as the only filler [13]. For correlating the structure and property of the shielding materials, trials were performed using nickel as it is conductive as well as ferromagnetic in nature. Various properties, such as hysteresis magnetic energy loss and magnetic coercive force [14], electrical resistivity, and shielding effectiveness [15], were studied on nickel filaments of varying diameters. It was found that as the diameter of the filaments increased and the coating was substituted with a nonconducting matrix, all of nickel's electrical and magnetic properties deteriorated.

Carbon-based materials, for example, coke, carbon fiber, and graphite, are conductive and have good radiation-absorbing capacity. Flexible graphite, a carbon-based shielding material, is used as EMI gasket material. It is also used for microelectronic applications. Because of skin effect and high surface area, nanocarbons such as carbon nanotubes and graphene have proven to be good shielding materials. Hybrid and porous nanocarbons are being used as shielding materials. In a hybrid composite, the materials can be carbon fiber, graphene, and nickel nanoparticles [16]. Metal-coated carbon fibers are very effective in shielding radiation, for example, nickel-coated carbon filament. A continuous carbon fiber polymer matrix and carbon-carbon composites are also effective shielding materials. Multiscale or hierarchical composites containing carbon fibers of nano and micro sizes are also being used in this field [7].

As ceramics have low conductivity, they are used for few applications in the EMI shielding field. Examples are silicon and titanium carbide, which show better conductivity than other ceramics. But magnetic ceramics such as ferrites show high absorption of radiation and are, therefore, used for shielding applications. But as these ceramics are nonconductive, they need to be combined with a carbon nanotube or reduced graphene oxide to impart conductivity. Commonly used ceramic grades are ceramic-carbon composites, for example, silver-coated ceramic particles [17]. Because they have high-temperature resistance, SiC-C composites are used for shielding purposes. A newer grade of ceramics, MXenes, that have enhanced properties have been discovered by researchers. To reduce costs, carbon fibers have been substituted by glass fibers in applications such as wind turbines and boats. To impart shielding properties, carbon nanotubes are added to the polymer matrix [18].

Normally, polymers are nonconductive and transparent to radiation. Metal-coated polymer fibers are said to have good shielding properties. Conducting polymers also are effective for this purpose [19]. Examples include polyaniline and polyacetylene. But they are very expensive and have poor mechanical properties. The matrix phase of composites is constituted primarily of the conducting polymers. The polymer used imparts good dispersibility to the filler and thereby helps to improve its shielding property.

Cement-based materials can act as effective shielding materials only when some admixtures are added to them. The admixtures comprise ingredients other than the regular ones that are added to enhance the properties of the cured cement. Examples of shielding enhancers are short fibers of carbon [20], carbon nanotubes [21], and steel. Some shielding enhancers are in the form of conductive particles, namely, coke [22], graphite, nickel, graphene, etc. Carbon-based materials are widely used here. Some ceramics are also being used, such as fly ash. The contact between the discontinuous particles to form a continuous conduction path enhances the shielding. For this purpose, the admixture should be added in a suitable quantity to the cement. The dispersion of these fibers or particles decreases as their particle size decreases. As the fiber volume fraction increases, the shielding effectiveness increases.

1.4 CONCLUSION

In today's electronic age with the rising demand for electronic devices, safety and damage prevention also need to be given importance. The discovery of EMI shielding materials has proven to be a boon for this. In days to come, more materials satisfying the previously mentioned needs will be explored and life will become smoother and easier. In the forthcoming chapters, more information on these shielding materials is given.

REFERENCES

1. S. Geetha, K.K. Satheesh Kumar, Chepuri R.K. Rao, M. Vijayan, D.C. Trivedi, EMI shielding: Methods and materials-a review, *J. Appl. Poly. Sci.*, 112, 2009, 2073–2086.
2. M. Mordiguine, *Interference Control in Computer and Microprocessors Based Equipments,* Don White Consultant Inc: Warrenton, 1984.
3. J.L.N. Violette, D.R.J. White, M.F. Violette, *Electromagnetic Compatibility Handbook*, Van Nostrand Reinhold Company: New York, 1987.
4. J.O. Jang, J.W. Park, U.S. Pat 6, 355, 707 B1 2002.
5. H.K. Miller, The EMF controversy: Are Electromagnetic fields dangerous to your health? *Mater. Eval.*, 55, 1997, 994.
6. Z. Sienkiewicz, Biological effects of electromagnetic fields, *Power Eng. J.*, 12, 1998, 131.
7. D.D.L. Chung, Materials for EMI shielding, *Materials Chem. & Phy.*, 255, 2020, 123587.
8. Y. Du, T. Liu, B. Yu, H. Gao, P. Xu, J. Wang, et al., The electromagnetic properties and microwave absorption of mesoporous carbon, *Mater. Chem. & Phy.*, 135, 2012, 884–891.

9. D. Zhang, T. Liu, J. Shu, S. Liang, X. Wang, J. Cheng, et al., Self-assembly construction of WS-rGO architecture with green EMI shielding, *ACS Appl. Mater. Interfaces*, 11, 30, 2019, 26807–26816.
10. Y.Y. Wang, X. Jing, Intrinsically conducting polymers for electromagnetic interference shielding, *Polym. Adv. Technol.*, 16, 2005, 344.
11. M. Cao, X. Wang, W. Cao, X. Fang, B. Wen, J. Yuan, Thermally driven transport and relaxation switching self-powered electromagnetic energy conversion, *Small*, 14, 29, 2018, 1800987.
12. A.V. Menon, G. Madras, S. Bose, Magnetic alloy-MWNT heterostructure as efficient electromagnetic wave suppressors in soft nanocomposites, *ChemistrySelect*, 2, 26, 2017, 7831–7844.
13. J. Wu, D.D.L. Chung, Combined use of magnetic and electrically conductive fillers in a polymer matrix for EMI shielding, *J. Electron Mater.*, 37, 8, 2008, 1088–1094.
14. X. Shui, D.D.L. Chung, Magnetic properties of nickel filament polymer-matrix composites, *J. Electron. Mater.*, 25, 6, 1996, 930–934.
15. X. Shui, D.D.L. Chung, Nickel filament polymer-matrix composites with low surface and high EMI shielding effectiveness, *J. Elect. Mater.*, 26,8, 1997, 928–934.
16. C. Wan, Y. Jiao, X. Li, W. Tian, J. Li, Y. Wu, A multi-dimensional and level by level assembly strategy for constructing flexible and sandwich-type nanoheterostructures for high-performance EMI shielding, *Nanoscale*, 12, 5, 2020, 3308–3316.
17. K.N. Tousif, Variation of electrical characteristics of EMI shielding materials coated with ceramic microspheres, *Int. Journal of Science Eng. and Tech.*, 1, 5, 2012, 185–188.
18. J.A. Rojas, B. Ribeiro, M.C. Rezende, Influence of serrated edge and rectangular strips of MWCNT buckypaper on the electromagnetic properties of glass fiber/epoxy resin composites, *Carbon*, 160, 2020, 317–327.
19. E. Hosseini, M. Arjmand, U. Sundararaj, K. Karan, Filler-free conducting polymers as a new class of transparent EMI shields, *ACS Appl. Mater. Interfaces*, 12, 25, 2020, 28596–28606.
20. S. Muthusamy, D.D.L. Chung, Carbon fiber cement-based materials for electromagnetic interference shielding, *ACI Mater. J.*, 107, 6, 2010, 602–610.
21. D. Micheli, A. Vricella, R. Pastore, A. Delfini, M.R. Bueno, M. Marchetti, F. Santoni, L. Bastianelli, F. Moglie, V. Mariani Primiani, et al., Electromagnetic properties of carbon nanotube reinforced concrete composites for frequency selective shielding structures, *Constr. Build. Mater.*, 131, 2017, 267–277.
22. S. Huang, G. Chen, Q. Luo, Y. Xu, Electromagnetic shielding effectiveness of carbon black-carbon fiber cement based materials, *Adv. Mater. Res.*, 168–170, 2011, 1438–1442.

2

CARBON MATERIALS

Potential Agents in Electromagnetic Interference Shielding

Suji Mary Zachariah, P. A. Nizam, Yves Grohens, Nandakumar Kalarikkal, and Sabu Thomas

⇒ **CONTENTS**

2.1 Introduction 9
2.2 Factors Affecting EMI SE 11
2.3 Advanced Carbon Hybrids for EMI Shielding 11
2.3.1 Foams 12
2.3.2 Aerogels 13
2.3.3 Films 14
2.3.4 Sandwiched Structures 15
2.3.5 Segregated Structures 16
2.3.6 Core-Shell Structures 17
2.3.7 Honeycomb Structures 18
2.4 Conclusion 18
References 19

2.1 INTRODUCTION

The demand for electromagnetic radiation (EMR) shielding materials has increased significantly in the past decade due to the rapid advancements in high-technology sectors like aerospace, electronics, and telecommunications. The shielding process basically consists of protecting a given component from electromagnetic waves (EMWs) by using enclosures that are made of electrically conductive or magnetic materials. As electrical conductivity (σ) is a major criterion for attenuating the EMWs, metal sheets or screens made of steel, copper, nickel, or aluminum alloys were used to a large extent. But, high density, poor corrosion resistance, cost of processing, and lack

DOI: 10.1201/9781003217312-2

of structural flexibility limited their usefulness in EMR shielding. Additionally, metal-based shielding materials reflect a major portion of the incoming EMR, causing secondary pollution that can cause damage to other electronic circuits or components. Few of these limitations could be resolved using intrinsically conducting polymers (ICP), but low thermal stability, mechanical performance, and cost remain unchanged. Hence, polymer composites with carbonaceous fillers have been adopted as a better alternative, as they impart many multifunctional characteristics like high surface area, improved σ values, and mechanical, thermal, and electrical properties.

Nano sized carbonaceous fillers, such as colloidal graphite, graphene (GN), graphene nanoplatelets (GNP), graphene oxide (GO), reduced graphene oxide (rGO), carbon nanotubes (CNTs), carbon fiber (CF), and carbon black (CB), have developed into one of the most promising components in electromagnetic shielding due to their high σ values, good mechanical performance, lightweight nature, flexibility features, and large aspect ratios. The geometry, size, dispersion, and distribution of filler particles contribute toward the fabrication of better EMR shielding materials. However, a lower filler content is usually preferred to reduce cost and density and for ease of processability. As uniform filler dispersion is a major criterion, two approaches, namely, kinetic and thermodynamic, have been utilized so far. The former approach involves using ultrasonication, while the latter approach involves using chemical additives that increase polymer-filler adhesion[1].

The CB particles possess a chainlike aggregation structure, thereby forming a stronger conductive network than other carbon-based materials. Also, they possess high specific strength and conductivity, making them a common EMR shielding material. However, their absorption capability is very poor, which results in poor attenuation of EMWs when used as a single medium[2]. Therefore, CF is usually combined with dielectric materials to achieve better shielding value. CF is interlocked sheets of carbon atoms arranged in regular hexagonal patterns with good reinforcing capability, modulus, strength, and low-density features. GN is a new member of the carbon allotrope family and a prime constituent in many nanocomposites for various applications. It was first obtained by two scientists, Andre Geim and Konstantin Novoselov, in 2004 through a mechanical peeling technique, which retained exceptional electrical conductivity and mechanical properties[3]. Its 2D nature helps in constructing anisotropic composites with layered structures, where they possess different properties between the direction of alignment and that perpendicular to it. This property of graphene is very difficult to achieve using other forms of carbon. CNTs are curled GN sheets and include single-walled carbon nanotubes (SWCNTs) and multiwalled carbon nanotubes (MWCNTs)[4].

2.2 FACTORS AFFECTING EMI SE

Shielding effectiveness is the ability of a material to protect certain devices or instruments from EMR. An ideal material for shielding applications should possess certain features for mitigating the effects of EMR. First, impedance matching between free space (air) and the surface of the shielding material is required. This helps with wave propagation into the material, thereby preventing reflection. This feature is greatly influenced by permittivity and permeability. Materials with large permeability absorb more EMWs, which is possible using magnetic materials like ferrites. Also, high-permittivity materials like $BaTiO_3$, TiO_2, MnO_2, ZnO, ZrO_2, and SiO_2 are also employed for absorption-dominated shielding. Even though such absorption type shields are environmentally safe and most preferred, a sufficient SE value cannot be achieved without the use of suitable conductive materials. Conductive materials used include various forms of metals and carbon materials. Another important factor is the thickness. Higher SE values can be obtained by increasing the sample thickness, but this often becomes a serious limitation when density and cost are taken into account[1].

2.3 ADVANCED CARBON HYBRIDS FOR EMI SHIELDING

There are various types or forms of carbon-based electromagnetic interference (EMI) shielding materials available on the market, such as films, composites, foams, aerogels, and textiles. Films are well-known to be a good choice for EMI shielding because of their ultrathin plane structure, they are lightweight and flexible, and they have a simple preparation process. The composite attributes that have drawn the most attention are their high strength and modulus and corrosion resistance. Foam is widely used in electromagnetic shielding to ascribe lightness and a large surface area. Due to the softness, flexibility, breathability, low cost, and simple preparation of textiles, they have developed into a research hots pot for electromagnetic shielding materials. Herein, we summarize the electromagnetic shielding properties of some carbon-based materials in terms of their performance, structure, type, and preparation.

2.3.1 Foams

A foam is a material in which air is trapped. It offers numerous advantages in shielding applications. First, the weight of the material can be considerably reduced, which is particularly important in aerospace applications and the next generation of portable electronic devices where weight is a crucial parameter. Second, even if the carbon filler is diluted in volume, the concentration of particles within the cell walls of the foam keeps the average distance between them almost the same, which is highly desirable when high electrical conductivity is desired at low carbon filler loading. Third, the presence of air inside the material decreases the real part of the permittivity, consequently reducing the reflectivity at the material's surface[1,4].

Ameli et al.[5] fabricated PP-CF foamed composites using an injection molding process and nitrogen gas as the foaming agent. Analysis revealed that the foaming process changed the microstructural configuration of the composites by a biaxial stretching effect on the fibers resulting in cell growth, increased the fibers' interconnectivity and orientation in the thickness direction, reduced fiber breakage, and decreased the size of the skin region. Finally, all of these improvements were reflected in the lowering of the percolation threshold from 8.75 to 7 vol% CF, enhancing the conductivity, dielectric permittivity, and SE value.

Renewable source–derived carbon foams and graphene have attracted extensive attention due to their 3D porous structure and remarkable electrical conductivity. Keeping this in mind, Wang et al.[6] prepared annealed sugarcane (ASC) by removing lignin from the sugarcane via a hydrothermal reaction followed by an annealing treatment. Later, ASC was filled with GO by vacuum-assisted impregnation and then thermally annealed to obtain the ASC/rGO hybrid foam. The synergistic effect of ASC and rGO resulted in attenuation of EMWs by virtue of their unique porous structures and abundant interfaces. The ASC/rGO foam with 17 wt% rGO content and SE value of 53 dB was obtained at a density of 0.047g/cm^3 and conductivity of 6 S/cm. Furthermore, all of these values were comparatively higher than those reported for ASC, with excellent flame retardancy, thermal stability, and heat insulation properties.

CNT/chitosan (CS) foams were assembled by Li et al.[7] using the freeze-drying method, giving excellent EMI shielding performance as well as outstanding mechanical performance. The foam exhibited an SE value of 37.6 dB at a density of 17.6 mg/cm^3, with absorption coefficient (*A*) and specific SE (SSE) values of 81.73% and 8556 dB·cm^2/g, respectively, which are attributed to the perfectly conductive networks. More importantly, the addition of CS significantly increased the compressive strength and modulus of CNT/CS foam to 34.1 kPa and 177.1 kPa, which were 84% and 149% higher than those for the pure CNT foam, respectively.

2.3.2 Aerogels

Aerogels are synthetic, ultralight materials having a porous, 3D solid network with air pockets. Due to their unique features like large open pores, high internal surface area, and low density, the infusion of conductive and magnetic fillers into aerogels is a promising means to prepare lightweight EMI shielding materials[8–10].

Narayanan et al.[11] demonstrated that V_2O_5, along with conducting polymer PANI, could be used for generating robust, green, EMI shielding aerogels with a good shielding performance value of 34.74 dB in the X-band. Interestingly, the SE value of the V_2O_5/PANI composite shows a 50% increase from that of the sample without polymer, which is confirmed by the electrical conductivity value that increased from 2.6×10^{-3} to 1.69×10^{-2} S/cm and the surface area of the mesoporous samples. Additionally, the 3D porous network structure made the density of the aerogel as low as 0.02 g/cm^3 and, hence, the SSE value of the aerogel achieved a high value of 1,662.2 dB cm^3/g.

Zhou et al.[12] developed a highly porous and conductive CNT-based carbon aerogel by two-stage pyrolysis and potassium hydroxide activation processes. The resultant activated, cellulose-derived carbon aerogels (a-CCAs) showed an ultrahigh EMI shielding performance of 96.4 dB and a high absorption coefficient of 0.79 in the X-band at a density of 30.5 mg/cm^3. These features are attributed to the construction of a hierarchically porous structure of a-CCAs and the introduction of a CNT-based heterogeneous conductive network that effectively dissipated the incident EMWs by interfacial polarization and microcurrent losses. The structure proposed here provided a possible pathway to overcome the conflict between high EMI shielding performance and ultralow or no secondary reflection.

Bi et al.[13] extensively studied the shielding performance of chemically (by hydrazine) and thermally reduced graphene aerogels (GAC and GAT) and obtained 27.6 and 40.2 dB at 2.5-mm thickness, respectively. Results revealed that the introduction of N_2 atoms through chemical reduction induced localized charges on the carbon backbone, leading to strong polarization effects. Also, incomplete reduction prevented the graphene sheets from p-p stacking. The higher extent of reduction of graphene sheets in GAT left only a few polar side groups, formed more sp2 graphitic lattice, and favored p-p stacking between the adjacent graphene sheets, resulting in higher electrical conductivity and a better SE value than for GAC.

2.3.3 Films

Films are obviously a good option for shielding applications due to their ultrathin plane structure, light weight, and simple preparation process. To date, several methods of preparing thin film have been developed, including vacuum filtration, microwave irradiation, centrifugal evaporation, electrophoretic deposition, chemical vapor deposition (CVD), electrospinning, solution casting, and layer-by-layer (LbL) assembly[14–20].

Shen et al.[21] reported the fabrication of GO films by direct evaporation of a GO suspension under mild heating. The resulting graphene film exhibited a graphite-like structure with an SE value of 20 dB at 8.4 μm and an in-plane thermal conductivity of 1100 W/m·K. Also, the GO film showed excellent mechanical flexibility and integrity during bending, indicating that the graphitization of GO film could be considered as a new, alternative way to produce excellent thermally conducting materials with better EMI shielding.

Zhou et al.[22] studied the synergistic effect of carbon nanotube (CNT) and graphene (GN) cast on a PMMA substrate. Analysis revealed that at 20 wt% CNT loading, the electrical conductivity value increased from 1.78×10^5 to 2.74×10^5 S/m, thermal conductivity increased from 510 to 1,154 W/m·K, and the SE value increased from 50 to 60 dB in graphene/CNT composite film (GCF). The reason for such a synergistic effect is due to the densification and bridge effects of the CNTs present in GCF.

In their work, Li et al.[23] developed two different flexible, free-standing CNT/PANI films, namely, unmodified (U-CNT/PANI) and amine functionalized (A-CNT/PANI) films. The A-CNT/PANI films reported SSE and σ values greater than those of U-CNT/PANI films. An SSE of 7.5×10^4 dB·cm^2/g and σ value of 3,009 S/cm were reported for the A-CNT/PANI film. Fourier transform infrared spectroscopy (FTIR) analysis clearly depicted the polymerization of NH_2 on the surface of CNT and aniline, which resulted in high conductivity in the composite system. Morphological analysis revealed that NH_2 groups on the CNT surface impart better wettability to A-CNT, resulting in uniform deposition of aniline on the A-CN and better conductivity. As conductivity is directly related to shielding effectiveness, the A-CNT/PANI film achieved a higher EMI SE of 50.2 dB than that of U-CNT/PANI because of its higher conductivity.

2.3.4 Sandwiched Structures

It has been an uphill task to attempt to overcome the challenge of incorporating superior folding resistance and high EMI shielding performance together into a carbon-based material with an interconnected, porous layered structure. Recently, Fu et al.[24] developed a special novel structure from SWCNT and graphene film (SGF), with an interconnected, porous, layered sandwich structure, via welding SWCNTs between graphene layers as the skeleton. The structure imparted SGF with folding-resistant performance under repeated folding over 1,000 times, without any variations in structural, mechanical, or electrical properties. At the same time, an exceptional SE value of 80 dB was achieved due to the conductivity of SWCNTs and the graphene network, as well as multiple reflections built up in the interlayers of SGF.

Singh et al.[25] developed an ingenious lightweight composite with an SE value greater than 37 dB in the Ku-band (12.4–18 GHz). The novel sandwich structure consists of iron oxide particles infiltrated in vertically aligned, highly porous CNTs, which are sandwiched along the tubular axis by rGO sheets. The excellent shielding performance is attributed to the presence of iron nanoparticles (NPs) in between the CNTs and residual defects or groups in rGO.

Wang et al.[26] fabricated hydrosensitive sandwich structures with self-tunable capability by sandwiching electrically conductive, pyrolytic graphite papers into polymeric, porous carbon-based spacers. By increasing the water loading of the spacers, the electromagnetic response capability was substantially improved via modifying the polarization and loss in the spacers. As a result, the shielding performance of the sandwich structures greatly changed upon different water loading values, indicating self-tunable EMI shielding performance.

2.3.5 Segregated Structures

Segregated structures offer numerous advantages in EMI shielding as they improve electrical conductivity and shielding effectiveness simultaneously. In such structures or architectures, electrical nanofillers are distributed at the interface of polymer granules rather than homogeneously distributing in the polymer matrix. But, the presence of nanofiller agglomerates at the interfaces restricts the diffusion of polymer granules, leading to the weak mechanical performance of segregated materials[1].

Limited study has been conducted on the effects of the synergistic combination of hybrid fillers and segregated structures on shielding performance. Jia et al.[27] extensively studied the synergistic effect of graphite and CNT (G-CNT) in ultrahigh molecular weight polyethylene (UHMWPE) composites, where the G-CNT hybrid was selectively distributed at the interfaces of UHMWPE domains. The resultant composite (G-CNT/UHMWPE) showed an excellent σ value of 195.3 S/m and an impressive SE value of 81.0 dB, values that were found to be superior to those of graphite alone or the CNT-loaded one, confirming the synergistic effect of graphite and CNT.

Zhan et al.[28] developed NR/Fe_3O_4@rGO (Natural rubber/ferrite2reduced graphene oxide (NRMG)) composites with segregated structures that have excellent SE values and stability under the bending cycle using a self-assembly and compression molding process. The NRMG composite showed a higher SE value of 42.4 dB than did the NR/rGO composite (NRG) of 34.0 dB at 10 phr rGO loading in the X-band frequency range. The excellent magnetic property imparted by Fe_3O_4 NPs on rGO sheets, together with good σ values, is attributed to the outstanding shielding performance of NRMG composites.

An reduced graphene oxide/polystyrene (rGO/PS) composite with a superior SE value, low filler loading, and enhanced mechanical performance has been developed through a combination of segregated architecture and high-pressure, solid-phase compression molding by Yan et al.[29]. An SE value of 45.1 dB is achieved with only 3.47 vol% rGO loading due to the segregated architecture that provides many interfaces to absorb the EM waves. This special architecture also reduced the amount of rGO by confining it at the interfaces and dramatically enhanced the mechanical strength by using a high pressure of 350 MPa, overcoming the major disadvantage of the composite made using conventional pressure (5 MPa). This also showed a 94% and

40% increase in compressive strength and compressive modulus, respectively.

2.3.6 Core-Shell Structures

Magnetic core-shell structured composites belong to the class of functional composites that has attracted huge attention for their novel performance in numerous applications. The shell and the core parts impart different physical and chemical characteristics and show significant EM shielding properties due to their peculiar morphology. The core-shell structure consists of core nanoparticles encapsulated inside the shell.

Yuan et al.[30] fabricated a free-standing composite film with a combined core-shell and sandwich microstructure by polymerizing polypyrrole onto $Fe_3O_4@SiO_2$ nonwoven fabrics, followed by an rGO coating (FSPG film). The film displayed SE and SSE values of 32 dB and 12,608.4 dB·cm^2/g, respectively, at 0.27-mm thickness. This high performance is due to the high σ value of the film (0.71 S/cm) and multiple internal reflections from the interconnected core-shell and sandwich microstructure. Additionally, the film, due to its unique interconnected backbone, has excellent mechanical flexibility and hence can sustain repeated bending and buckling to nearly 180 degrees.

CFs encapsulated with nano-Cu with excellent hydrophobicity, antibacterial activity, and EM shielding properties have been reported. The multipurpose, core-shell structured composite demonstrated a high σ value, contributing to an SE value of 29 dB in the X-band frequency range. It was also reported that the absorption-dominant mechanism is dominant due to the higher conductivity and magnetic permeability of the material, which are therefore beneficial for alleviating secondary radiation[31].

A flexible, lightweight, microwave-absorbing material based on PVDF nanofibers with a poly(3,4-ethylenedioxythiophene) (PEDOT) polymerized shell has been fabricated by vapor phase polymerization (VPP) of ethylenedioxythiophene (EDOT) monomers. It was reported that, prior to the VPP process, the PVDF nanofibers show SE values of less than 5 dB in the X-band frequency region due to their lack of conductivity. But after the VPP process, an SE value of 40 dB was reported at 16 wt% PVDF nanofiber content. Meanwhile, the effect of doping with iron chloride on the SE value was studied at a constant PVDF concentration. The results indicate that the SE value, along with other features like durability and flexibility, is highest at 4 wt% doping[32].

2.3.7 Honeycomb Structures

Honeycomb-type architecture is the most advantageous structure for making EM shielding materials, compared with porous 3D structures. As the EM waves enter the structure, they undergo scattering, absorption, and multiple internal reflections, which prolong the waves' residence time inside the structure and consequently enhance its shielding ability. Also, the filler dispersion is uniform in honeycomb structures, the geometry is easy to control, and they can reduce the weight of the final composite.

In their work, Zeng et al.[33] fabricated a robust and environmentally benign EM shielding material from lignin, which is an abundant biomass material that accounts for about 35% of the content of wood. The honeycomb-like, lignin-derived carbon foam (LC) doped with rGO was fabricated by a unidirectional ice-template method followed by freeze drying and carbonization. The authors revealed that the sample could attain 70 dB SE because of its high σ value, aligned pores, and numerous interfaces existing between LC and rGO. Also, its dimensions, composition, and density could be easily tuned in order to achieve high σ values and mechanical properties

Song et al.[34] prepared an rGH (rGO with a honeycomb structure)/epoxy composite with excellent σ and SE values by simple mixing and curing. An Al_2O_3 honeycomb was used as a template for making rGH, which was later freeze dried and thermally annealed so as to be immersed in epoxy resin to form the composite. Interestingly, the rGH-incorporated composite showed better conductivity and a higher SE value than the composite with rGO at the same wt%. The epoxy/rGH composite showed an SE value and a σ value of 38 dB and 40.2 S/m, respectively, whereas for the epoxy/rGO composite they were 6 dB and 0.03 S/m, respectively. The better SE value for the rGH/epoxy composite is attributed to the 3D honeycomb structure of rGH, where the EMWs are reflected and scattered multiple times, resulting in prolonged propagation paths and, finally, better shielding efficiency.

2.4 CONCLUSION

This chapter reviewed numerous advanced carbon-based materials for EM shielding applications with a particular emphasis on novel structure/design strategies like aerogels, foams, honeycomb structures, and segregated structures. Even though the past few decades have witnessed various versatile materials that substitute for metals due to their superior shielding ability, the cost factor and a greener approach to fabrication have been ignored in most cases. For instance, the most investigated materials in the past few years are the MXenes, due to their excellent SE values and mechanical flexibility, but they face the limitations of a complex preparation process and cost. Furthermore, it is also challenging in many cases to achieve the uniform dispersion of fillers while maintaining good σ values. Even if the filler content is reduced, uniform filler dispersion cannot be achieved. On the other hand, when the filler content is increased, very high σ values can be readily obtained, but this results in an increase in the total weight of the composites and it reduces flexibility. Hence, there is still a long path ahead to develop shielding materials as the technology will keep on growing in the coming years, and, once all of these challenges are met, the aim to protect devices from stray radiation will be realized.

REFERENCES

1. Jaroszewski, M, Thomas, S, and Rane, AV (Eds.). *Advanced Materials for Electromagnetic Shielding: Fundamentals, Properties, and Applications*, pp. 147–166. Hoboken, NJ: John Wiley & Sons, 2018.
2. Tarasova, E, Byzova, A, Savest, N, et al. "Influence of preparation process on morphology and conductivity of carbon black-based electrospun nanofibers." *Fullerene Nanotube Carbon Nanostructures* 23 (2014): 695–700.
3. Geim, AK, and Novoselov, KS. The rise of graphene. *Nature Materials* 6 (2007): 183–191.
4. Thomassin, J-M, Je´ro^me, C, Pardoen, T, et al. "Polymer/ carbon based composites as electromagnetic interference (EMI) shielding materials." *Materials Science and Engineering R: Reports* 74 (2013): 211–232.
5. Ameli, A, Jung, PU, and Park, CB. "Electrical properties and electromagnetic interference shielding effectiveness of polypropylene/ carbon fiber composite foams." *Carbon* 60 (2013): 379–391.
6. Wang, Lei, et al. "Lightweight and robust rGO/sugarcane derived hybrid carbon foams with outstanding EMI shielding performance." *Journal of Materials Science & Technology* 52 (2020): 119–126.
7. Li, Meng-Zhu, et al. "Robust carbon nanotube foam for efficient electromagnetic interference shielding and microwave absorption." *Journal of Colloid and Interface Science* 530 (2018): 113–119.
8. Long, Jeffrey W., et al. "Three-dimensional battery architectures." *Chemical Reviews* 104.10 (2004): 4463–4492.

9. Huang, Hua-Dong, et al. "Cellulose composite aerogel for highly efficient electromagnetic interference shielding." *Journal of Materials Chemistry A* 3.9 (2015): 4983–4991.
10. Chen, Yian, et al. "Multifunctional cellulose/rGO/Fe3O4 composite aerogels for electromagnetic interference shielding." *ACS Applied Materials & Interfaces* 12.19 (2020): 22088–22098.
11. Narayanan, Aparna Puthiyedath, Narayanan Unni, KN, and Peethambharan Surendran, Kuzhichalil. "Aerogels of V2O5 nanowires reinforced by polyaniline for electromagnetic interference shielding." *Chemical Engineering Journal* 408 (2021): 127239.
12. Zhou, Zi-Han, et al. "Structuring hierarchically porous architecture in biomass-derived carbon aerogels for simultaneously achieving high electromagnetic interference shielding effectiveness and high absorption coefficient." *ACS Applied Materials & Interfaces* 12.16 (2020): 18840–18849.
13. Bi, Shuguang, et al. "A comparative study on electromagnetic interference shielding behaviors of chemically reduced and thermally reduced graphene aerogels." *Journal of Colloid and Interface Science* 492 (2017): 112–118.
14. Xing, D, Lu, L, Xie, Y, et al. "Highly flexible and ultra-thin carbon-fabric/Ag/waterborne polyurethane film for ultra-efficient EMI shielding." *Materials & Design* 185 (2020): 108227.
15. Kumar, P, Shahzad, F, Yu, S, et al. "Large-area reduced graphene oxide thin film with excellent thermal conductivity and electromagnetic interference shielding effectiveness." *Carbon* 94 (2015): 494–500.
16. Lee, S-H, Kang, D, and Oh, KI. "Multilayered graphenecarbon nanotube-iron oxide three-dimensional heterostructure for flexible electromagnetic interference shielding film." *Carbon* 11 (2017): 248–257.
17. Shen, B, Zhai, W, and Zheng, W. "Ultrathin flexible graphene film: An excellent thermal conducting material with efficient EMI shielding." *Advanced Functional Materials* 24 (2014): 4542–4548.
18. Song, WL, Gong, C, Li, H, et al. "Graphene-based sandwich structures for frequency selectable electromagnetic shielding." *ACS Applied Materials & Interfaces* 9 (2017): 36119–36129.
19. Wang, Z, Mao, B, Wang, Q, et al. "Ultrahigh conductive copper/large flake size graphene heterostructure thin-film with remarkable electromagnetic interference shielding effectiveness." *Small* 14 (2018): e1704332.
20. Wu, ZP, Cheng, DM, Ma, WJ, et al. "Electromagnetic interference shielding effectiveness of composite carbon nanotube macro-film at a high frequency range of 40 GHz to 60 GHz." *AIP Advances* 5 (2015): 067130.

21. Shen, Bin, Zhai, Wentao, and Zheng, Wenge. "Ultrathin flexible graphene film: An excellent thermal conducting material with efficient EMI shielding." *Advanced Functional Materials* 24.28 (2014): 4542–4548.
22. Zhou, Erzhen, et al. "Synergistic effect of graphene and carbon nanotube for high-performance electromagnetic interference shielding films." *Carbon* 133 (2018): 316–322.
23. Li, Hui, et al. "Lightweight flexible carbon nanotube/polyaniline films with outstanding EMI shielding properties." *Journal of Materials Chemistry C* 5.34 (2017): 8694–8698.
24. Fu, Huili, et al. "SWCNT-modulated folding-resistant sandwich-structured graphene film for high-performance electromagnetic interference shielding." *Carbon* 162 (2020): 490–496.
25. Singh, Avanish Pratap, et al. "Probing the engineered sandwich network of vertically aligned carbon nanotube—reduced graphene oxide composites for high performance electromagnetic interference shielding applications." *Carbon* 85 (2015): 79–88.
26. Wang, Yana, et al. "Hydro-sensitive sandwich structures for self-tunable smart electromagnetic shielding." *Chemical Engineering Journal* 344 (2018): 342–352.
27. Jia, Li-Chuan, et al. "Synergistic effect of graphite and carbon nanotubes on improved electromagnetic interference shielding performance in segregated composites." *Industrial & Engineering Chemistry Research* 57.35 (2018): 11929–11938.
28. Zhan, Yanhu, et al. "Fabrication of a flexible electromagnetic interference shielding Fe3O4@ reduced graphene oxide/natural rubber composite with segregated network." *Chemical Engineering Journal* 344 (2018): 184–193.
29. Yan, Ding-Xiang, et al. "Structured reduced graphene oxide/polymer composites for ultra-efficient electromagnetic interference shielding." *Advanced Functional Materials* 25.4 (2015): 559–566.
30. Yuan, Ye, et al. "Lightweight, flexible and strong core-shell non-woven fabrics covered by reduced graphene oxide for high-performance electromagnetic interference shielding." *Carbon* 130 (2018): 59–68.
31. Jiao, Yue, et al. "Carbon fibers encapsulated with nano-copper: A core—shell structured composite for antibacterial and electromagnetic interference shielding applications." *Nanomaterials* 9.3 (2019): 460.
32. Lee, Sol, et al. "Polyvinylidene fluoride core—shell nanofiber membranes with highly conductive shells for electromagnetic interference shielding." *ACS Applied Materials & Interfaces* 13 (2021): 25428–25437.

33. Zeng, Zhihui, et al. "Biomass-based honeycomb-like architectures for preparation of robust carbon foams with high electromagnetic interference shielding performance." *Carbon* 140 (2018): 227–236.
34. Song, Ping, et al. "Obviously improved electromagnetic interference shielding performances for epoxy composites via constructing honeycomb structural reduced graphene oxide." *Composites Science and Technology* 181 (2019): 107698.

3

NOVEL MXENE MATERIALS FOR ELECTROMAGNETIC INTERFERENCE SHIELDING APPLICATIONS

P. A. Nizam, Suji Mary Zachariah, and Sabu Thomas

⇒ CONTENTS

3.1 Introduction 23
3.2 History of EMI Shielding Materials 24
3.3 Common Fillers Used in Shielding Applications 25
3.4 Why MXenes for EMI Shielding? 26
3.5 Synthesis and Characterization of MXenes 27
3.6 Case Studies 29
3.7 Challenges 30
3.8 Conclusion 31
References 31

3.1 INTRODUCTION

Industrialization and technological evolution have escalated rapidly, compromising many aspects that directly or indirectly affect humans as well as nature. The advancements in the electronic sector[1] have had drastic effects as they produce electromagnetic pollution. Electromagnetic waves are the result when an electric field and a magnetic field come in contact with each other in a perpendicular direction. These waves move with a steady velocity of 2.998×10^8 m/s, deflected by no electric nor magnetic field. These waves propagate through media such as air, vacuum, and solid matter in the form of photons or quanta. Even though these waves do not deflect, they can interfere and diffract[2].

DOI: 10.1201/9781003217312-3

Electromagnetic interference (EMI) is the interference of the electromagnetic fields established by one electrical or electronic unit. There are several types of electromagnetic interference that may impair and hinder circuits from operating in the manner expected. EMI has been a major crisis in the fields of the military, aerospace, communications, and many other areas. It engenders a serious impact on the output of the circuit, which results in disastrous outcomes of induction and radiation from external sources[3]. These major problems in the field of electronics have spurred scientists to develop materials for EMI shielding that range from electronic systems to biological systems[4].

There has been a decrease in the usage of traditional materials such as metals and their composites, as they possess the disadvantages of high density, corrosiveness, flexibility, and processing cost. These parameters are overcome by materials such as polymers, carbon matrices, and other ceramic composites. Polymer nanocomposites play a vital role, as their shielding efficiency can be tailored by diversifying the composition of the filler[5]. Carbon-based filler systems like graphene or carbon nanotubes attracted immense attention in this field due to their excellent electrical conductivity, high aspect ratio, and light weight, but the achievement of excellent EMI shielding efficiency at smaller thicknesses is still a major challenge[6].

Recently, a new class of 2D transition carbides, nitrides, and carbonitrides with a metallically conductive and hydrophilic nature was discovered by Gogotsi and co-workers. They exhibit a higher efficiency in EMI shielding than most of the synthetic materials available, with a lower thickness. The material is promising for future applications in the fields of energy storage, water purification, and antibacterial properties.

3.2 HISTORY OF EMI SHIELDING MATERIALS

Electromagnetic radiation has been a major problem since the invention of electronic devices. The research in polymers for EMI shielding was very active by the 1970s, where conducting polymer plastics with the incorporation of metallic fillers were investigated[7]. Brass, nickel, silver, stainless steel, and metalized plastics were the main contributors to their conductivity, but these face many drawbacks. The high density of stainless steel and the low impact strength of aluminum lower their use in such applications. Other classes such as metallic glasses were researched, which also face the disadvantage of their brittle nature. Graphite was used in the early 1980s but was limited due to its high costs[8].

Intrinsically conducting polymers (ICPs) gained popularity as they evinced high shielding efficiency when incorporated with thermoplastic blends[9]. Several studies on ICPs have been performed, mainly using polyaniline and polypyrrole as the active filler and polymers such as epoxy, polyvinyl chloride, or polyvinyl alcohol as matrices[10]. By 1990, researchers began to investigate ICPs coated with electrically insulating materials for applications in EMI shielding, as they promised a weight reduction compared to other metal-based traditional composites[11].

In more recent years, carbon-based nanomaterials emerged as the best fillers for EMI shielding materials with the advantages of having light weight, high aspect ratio, electrical conductivity, and thermal stability. Multiwall carbon nanotubes show excellent shielding efficiency when compared to other fillers, such as carbon nanofibers and carbon blacks. Nanotubes were in the top position in shielding applications until 2D graphene was invented in 2008[12]. Graphene is an excellent nanomaterial for EMI shielding due to its superior conductivity and strength. It is produced by the exfoliation of graphite; graphene is a single sheet of graphite.

MXenes, a class of inorganic, metal-based carbides, nitrides, and carbonitrides discovered in 2011, exhibit better properties than graphene in terms of conductivity, stability, and strength. Several other metal-based nanoparticles have also been synthesized, such as silver, gold, copper, and iron; these were used along with other conducting materials to improve the final properties for EMI shielding.

3.3 COMMON FILLERS USED IN SHIELDING APPLICATIONS

The history of EMI shielding has shed light on the fillers used for EMI shielding applications. Metals are good conductors of electricity, which contributes to their use as EMI shielding fillers. A highly permeable alloy of copper, chromium, nickel, and iron, commonly known as mu-metal, is an excellent agent for shielding applications. Other metal fillers such as aluminum, silver, stainless steel, nickel, and brass also enhance the EMI shielding properties of polymers. Aluminum, copper, or silver metal foils embedded in the polymer matrix provide a mode of grounding provision. Their degree of shielding is based on the foil's thickness, quantity, and the medium in which it is been dispersed. They exhibit a higher EMI shielding efficiency, while simultaneously facing disadvantages such as high weight and low impact resistance.

The incorporation of conducting polymers such as polyaniline and polypyrrole has gained immense attention due to their stability and flexibility. Flexible polyaniline reinforced with water-soluble polyvinyl chloride (PVC), cellulose, and other polymer matrices results in resistivity in the range of 60 to 1,000 Ω·cm. Polymerization of pyrrole in a nylon surface shows a resistivity in the range 200 Ω·cm to 20 kΩ·cm. Furthermore, many inorganic nanoparticles such as Ag, Cu, and Pd, when incorporated in these composites, impart a higher shielding efficiency. To exemplify, Ag nanoparticles incorporated into a polyvinylidene fluoride (PVDF)/graphite composite demonstrated a broadly improved dielectric constant and at the same time lower dielectric loss[13].

Carbon-based nanotubes, single-walled nanotubes (SWNTs), and multi-walled nanotubes (MWNTs) have emerged as excellent fillers in recent years, despite their hazardous nature. Many studies in the field of EMI shielding proved the nanotubes to be a superior recipient for the application. The Nobel Prize–winning material graphene is another carbon-based, strong, and highly conductive material, which is made from the exfoliation of graphite[14]. These materials have replaced nanotubes in many applications due to their eco-friendly nature and properties. Numerous studies have reported on these two materials in the field of EMI shielding. Other carbon-based materials, like blacks, have gained little attention in this field.

A new family of inorganic metal carbides and nitrites, known as MXenes, has attracted significant notice due to their high metallic conductivity, good dispersibility, and high hydrophilic stability. These materials are promising for many applications, such as energy storage and EMI shielding batteries.

3.4 WHY MXENES FOR EMI SHIELDING?

MXenes, discovered by Gogotsi and co-workers in 2011, belong to a new class of inorganic 2D materials made from metal carbides, nitrides, and carbonitrides. These layered materials are named MXenes because they are formed by etching a layer in MAX phases, and the suffix "ene" emphasizes their resemblance to graphene. MAX phases are a broad family (60+ members) of ternary transit layered hexagons of carbides, carbonitrides, and nitrides of $M_{n+1}AX_n$

composition, where M is an early transition metals (for example, Ti, V, Cr, or Nb), A represents a Group A element (for example, Al, Si, Sn, or In), X represents carbon and/or nitrogen, and $n = 1, 2,$ or 3. MXenes typically have excellent properties, such as high chemical stability and high electric conductivity, and environmentally favorable characteristics because of their special structures. To date, MXenes have been identified as promising materials[15] for semiconductors, EMI shielding, supercapacitors, and lithium-ion batteries.

Lightweight, layer-by-layer, porous, and segregated structure composites were developed using MXenes thanks to their tunable surface chemistry[16]. In miniature electronics where thin shielding is required, MXene-based composites show excellent shielding along with excellent mechanical strength. MXene laminates have a comparatively limited thickness for better EMI SE and can be manufactured on a nanoscale to fulfill commercial requirements. The porous foams and aerogels have good ultralight EMI protection at the cost of thickness, which are beneficial for aeronautical and military applications. MXenes have all the key features required for an effective EMI protective material—strong electric conductance, wide specific surfaces, lightweight, and, above all, easy operation.

3.5 SYNTHESIS AND CHARACTERIZATION OF MXENES

Many protocols are reported for the synthesis of MXenes. Each method produces different characteristic properties of MXenes that are suitable for different applications. The main ingredient in the synthesis of MXenes is the precursor. They are prepared from the MAX phase via diverse routes, compositions, raw materials, and sintering conditions. Hydrofluoric acid (HF, 49.5 wt%) is commonly used as the etchant. A common route of synthesis involves the slow addition of precursor to the etchant with continuous stirring at room temperature. After etching is complete, the powders are centrifuged to wash them enough to bring the pH down to approximately 6. The sediments are properly washed, filtered, and dried in an oven to obtain the final product. Different concentrations of etchant can be used by varying the time of reaction, say for 5% concentration 24 hours, while for 30% concentration the reaction is complete in 5 hours. This variable concentration synthesis route produces different MXenes, with different morphologies and appeareances[17]. Higher concentrations reveal a darker color than lower concentrations. The morphology of higher concentrations shows an accordion-like structure that fades with decreases in the concentration of etchant. This can be attributed to the excess gas that escapes during the exothermic reaction. Some other etchants, such as NH_4HF_2 nd HF are efficient for the preparation of Mxene, where different concentrations of LiF and HCL are being employed to produce HF.[18] The quality of HF changes with the concentration of both of the reactants, which in turn affects the quality of the MXene formed. The best method to produce minimal defects in MXenes is the minimally intensive layer delamination (MILD) process. This method is favorable when the application requires larger, high-quality flakes of particles. Figure 3.1 shows the morphologies of (a) Ti_3AlC_2 with its compact structure, (b) at 30%, (c) at 10% (d) 5 wt% etchant, (e) the NH_4HF_2 mode of synthesis, and (f) MILD synthesis. Images d–f show MXene lamellas with negligible openings.

Although the 5 wt% HF etches aluminum from MAX phase Ti_3AlC_2, an accordion-like component morphology is generally observed when the HF solution is 10 wt% at a minimum. The same morphology is observed in *in situ* HF formation methods. For applications where high electrical conductivity, good mechanical properties, larger flake size, and environmental stability are required, it is better to use the MILD process, whereas smaller, more defective particles are yielded with the HF etching method. MXene flakes can be characterized by using X-ray diffraction (XRD), scanning

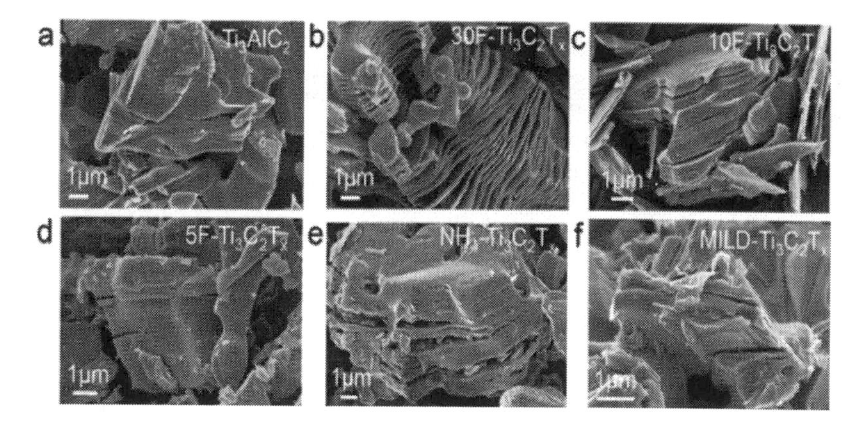

Figure 3.1 (a) Ti_3AlC_2 with its compact structure, (b) at 30%, (c) 10% (d) 5 wt% etchant; (e) the NH_4HF_2 mode of synthesis, and (f) MILD synthesis.

Source: Reprinted with permission from Alhabeb et al.[17]

electron microscope (SEM) and transmission electron microscope (TEM) and Raman spectroscopy for analysis of their structure and morphology. Elemental analysis and particle size are understood by energy dispersive x-ray analysis (EDX) and dynamic light scattering (DLS)[19].

3.6 CASE STUDIES

Many studies have been reported since the discovery of MXenes for EMI shielding. Different structural forms such as layer-by-layer assembly, laminate structure, aerogels, porous foams, and segregated structures have been researched on the basis of their density, thickness, and EMI SE. He et al. fabricated a green composite of an MXene and hydroxyethylcellulose via filtration-assisted self-assembly. The EM waves dissipated in the film in the form of heat, which was confirmed by dielectric loss analysis. The effect of multilayer stacking indicates that the film shows an absorption-dominant green EMI shielding[20]. Lightweight, absorption-dominant shielding materials are more enticing than their conventional reflection-dominant counterparts as they reduce double the amount of pollution from EM waves. Another study where a porous form of MXene/PVA was developed using the freeze-drying method exhibits specific shielding effectiveness of up to 5,136 dB·cm^2/g with just 0.15 vol%. The analysis reveals that superior EMI shielding and enhanced absorption efficiency are the outcomes of the internal reflection and good impedance derived from the highly porous structure[21]. Other than composites, robust MXene aerogels are made with tunable, aligned lamellar architecture and a compressible nature via bidirectional freeze-drying of the MXene solution. These anisotropic aerogels exhibit an excellent shielding efficiency of around 70.5 dB in the frequency range of 8.2–12.4 GHz. Moreover, after 20 compression-releasing cycles, they maintain a stress recovery of 84.5% and steady electrical resistance. The lightweight, moderate compression, additive-free MXene aerogels can be used to design the best resistance for lightweight and versatile electronic devices[22].

Many Chinese researchers have actively investigated MXenes and their various nanocomposites. An annealed, epoxy-based, MXene nanocomposite was prepared using the solution casting method, where the MXene was prepared via ionic intercalation and sonication. With a mass fraction of 15% annealed MXene, the nanocomposite exhibited an electrical conductivity of 105 S/m and shielding of 41 dB. These values are 176% and 37% higher than those of normal

MXene/epoxy nanocomposites (unannealed MXene), respectively. The sample also shows optimum Young's modulus and better hardness. Modified or annealed MXenes are much more efficient, as per this report[23]. Cao et al. fabricated flexible and ultrathin MXene/cellulose nanofiber (CNF) composite paper via a vacuum filtration–induced self-assembly method with a nacre-like microstructure. By adjusting the weight ratio of two components, a tensile strength of 135.4 MPa and a 16.7% strain fracture are achieved. A "brick and mortar" toughening process is indicated by the nacre-like layered structure in the MXene/CNF composite due to the close interaction of the 2D d-Ti_3C_2Tx (group A when itched is represented as Tx) nanosheet and 1D CNF[24]. The composite exhibited a high electrical conductivity and excellent EMI shielding and folding endurance, where folding numbers can reach as high as 14,260 times.

Numerous studies with different polymer matrices have yielded interesting results: Sun et al. prepared an MXene-loaded polystyrene nanocomposite via electrostatic assembly followed by compression molding. Negatively charged MXene nanosheets afford an excellent conducting network by electrostatically assembling onto positive microspheres of polystyrene. An electrical conductivity of 1,081 S/m and a low percolation threshold are achieved along with a 54 dB performance of EMI shielding via an absorption-dominant mechanism[25]. Graphene has been used in shielding applications due to its light weight and electrical conductivity. However, graphene oxide's inferior properties compared to graphene restrict its use in the application. MXene incorporated in GO foam via the freeze-drying method endows an excellent shielding material with high wave attenuation and shielding effectiveness of 50.7 dB. These hybrid composites with high specific shielding effectiveness (SSE) at a lower thickness have potential for future applications in next-generation smart devices[26]. The MXenes are still in their infancy, and there is a lot more to explore about these materials in the future.

3.7 CHALLENGES

Many technical issues have to be solved to fit this wonderful class of materials, MXenes, into useful applications. The synthesis of MXenes using hydrofluoric acid has severe health and environmental concerns. The employment of a less hazardous and green chemical will be accepted as a breakthrough in this field. Green synthesis is a major challenge for this wonderful material. The implementation of an effective, low-cost system for the mass manufacture of MXenes is another key challenge, which will enhance its commercialization. Another problem (and potential) for researchers to explore and overcome is the restriction of subzero temperatures for storing MXene dispersions. MXenes' thermal and oxidative stability in water is also a significant priority that researchers need to address, as these materials are thermodynamically metastable and oxidation resistance is moderate. More MXene structures need to be developed that possess these enhanced characteristics for different applications, particularly environmental remediation. Though MXene applications have been uncovered in various environmental fields, researchers remain incapable of incorporating MXenes into commercial applications. Many of MXenes' proposed environmental implementations are dependent on laboratory or experimental research. Although the real use of MXene-based materials is only an early step, the viability of real environmental applications must still be assessed in the current circumstances[27].

3.8 CONCLUSION

MXenes are the best contestants for potential applications to energy storage and EMI shielding batteries. They have exhibited excellent conductivity, thermal properties, and stability with enhanced mechanical properties, overtaking most of the other fillers that have been used in EMI shielding. The feasibility of their dispersal in matrices and their hydrophilic properties enhance their further use in applications. Tunable properties, based on their synthesis, morphology, and particle size, have been interesting topics discussion and need further investigation, as there are still platforms to explore. However, many challenges persist in this field that need to be addressed. These challenges can be overcome in the future, making MXenes excellent materials for many applications.

REFERENCES

1. Nizam, P. A. *et al. Nanocellulose-Based Composites. Nanocellulose Based Composites for Electronics* (Elsevier Inc., 2021). doi:10.1016/b978-0-12-822350-5.00002-3.
2. Pai, A. R., Paoloni, C. & Thomas, S. *Nanocellulose-Based Sustainable Microwave Absorbers to Stifle Electromagnetic Pollution. Nanocellulose Based Composites for Electronics* (Elsevier Inc., 2021). doi:10.1016/b978-0-12-822350-5.00010-2.

3. Gopakumar, D. A. *et al.* Flexible Papers Derived from Polypyrrole Deposited Cellulose Nanofibers for Enhanced Electromagnetic Interference Shielding in Gigahertz Frequencies. *J. Appl. Polym. Sci.* **138**, 1–11 (2021).
4. Pai, A. R. *et al.* Ultra-Fast Heat Dissipating Aerogels Derived from Polyaniline Anchored Cellulose Nanofibers as Sustainable Microwave Absorbers. *Carbohydr. Polym.* (2020). doi:10.1016/j.carbpol.2020.116663.
5. Gopakumar, D. A. *et al.* Cellulose Nanofiber-Based Polyaniline Flexible Papers as Sustainable Microwave Absorbers in the X-Band. *ACS Appl. Mater. Interfaces* **10**, 20032–20043 (2018).
6. Abbasi, H., Antunes, M. & Velasco, J. I. Recent Advances in Carbon-Based Polymer Nanocomposites for Electromagnetic Interference Shielding. *Prog. Mater. Sci.* **103**, 319–373 (2019).
7. Bigg, D. M. Conductive Polymeric. *Corn posi t ions**. **17**, (1977).
8. Simon, R. M. Emi Shielding Through Conductive Plastics. *Polym. Plast. Technol. Eng.* **17**, 1–10 (1981).
9. Colaneri, N. F. & Shacklette, L. W. EMI Shielding Measurements of Conductive Polymer Blends. *IEEE Trans. Instrum. Meas.* **41**, 291–297 (1992).
10. Mäkelä, T., Pienimaa, S., Taka, T., Jussila, S. & Isotalo, H. Thin Polyaniline Films in EMI Shielding. *Synth. Met.* **85**, 1335–1336 (1997).
11. Yang, Z., Peng, H., Wang, W. & Liu, T. Crystallization Behavior of Poly(ε-Caprolactone)/Layered Double Hydroxide Nanocomposites. *J. Appl. Polym. Sci.* **116**, 2658–2667 (2010).
12. Al-Saleh, M. H., Saadeh, W. H. & Sundararaj, U. EMI Shielding Effectiveness of Carbon Based Nanostructured Polymeric Materials: A Comparative Study. *Carbon N. Y.* **60**, 146–156 (2013).
13. Kumaran, R., Alagar, M., Dinesh Kumar, S., Subramanian, V. & Dinakaran, K. Ag Induced Electromagnetic Interference Shielding of Ag-Graphite/PVDF Flexible Nanocomposites Thinfilms. *Appl. Phys. Lett.* **107** (2015).
14. Nizam, P. A. *et al.* Mechanically Robust Antibacterial Nanopapers Through Mixed Dimensional Assembly for Anionic Dye Removal. *J. Polym. Environ.* **28**, 1279–1291 (2020).
15. Jiang, Q. *et al.* Review of MXene Electrochemical Micro-supercapacitors. *Energy Storage Mater.* (2020). doi:10.1016/j.ensm.2020.01.018.
16. Iqbal, A., Sambyal, P. & Koo, C. M. 2D MXenes for Electromagnetic Shielding: A Review. *Adv. Funct. Mater.* **30**, 1–25 (2020).
17. Alhabeb, M. *et al.* Guidelines for Synthesis and Processing of Two-Dimensional Titanium Carbide (Ti3C2Tx MXene). *Chem. Mater.* **29**, 7633–7644 (2017).

18. Shayesteh Zeraati, A. *et al.* Improved Synthesis of Ti3C2T: XMXenes Resulting in Exceptional Electrical Conductivity, High Synthesis Yield, and Enhanced Capacitance. *Nanoscale* (2021). doi:10.1039/d0nr06671k.
19. Shekhirev, M., Shuck, C. E., Sarycheva, A. & Gogotsi, Y. Characterization of MXenes at Every Step, from Their Precursors to Single Flakes and Assembled Films. *Prog. Mater. Sci.* (2020). doi:10.1016/j.pmatsci.2020.100757.
20. He, P. *et al.* Self-Assembling Flexible 2D Carbide MXene Film with Tunable Integrated Electron Migration and Group Relaxation Toward Energy Storage and Green EMI Shielding. *Carbon N. Y.* **157**, 80–89 (2020).
21. Xu, H. *et al.* Lightweight Ti 2 CT x MXene/Poly(vinyl alcohol) Composite Foams for Electromagnetic Wave Shielding with Absorption-Dominated Feature. *ACS Appl. Mater. Interfaces.* **11**, 10198–10207 (2019).
22. Han, M. *et al.* Anisotropic MXene Aerogels with a Mechanically Tunable Ratio of Electromagnetic Wave Reflection to Absorption. *Adv. Opt. Mater.* **7**, 1–7 (2019).
23. Wang, L. *et al.* Fabrication on the Annealed Ti 3 C 2 T x MXene/ Epoxy Nanocomposites for Electromagnetic Interference Shielding Application. *Compos. Part B.* **171**, 111–118 (2019).
24. Cao, W. *et al.* Binary Strengthening and Toughening of. *ACS Nano.* **12**, 4583–4593 (2018).
25. Sun, R. *et al.* Highly Conductive Transition Metal Carbide/ Carbonitride(MXene)@polystyrene Nanocomposites Fabricated by Electrostatic Assembly for Highly Efficient Electromagnetic Interference Shielding. *Adv. Funct. Mater.* **27**, 1–11 (2017).
26. Fan, Z. *et al.* A Lightweight and Conductive MXene/Graphene Hybrid Foam for Superior Electromagnetic Interference Shielding. *Chem. Eng. J.* **381**, 122696 (2020).
27. Ihsanullah, I. & Ali, H. Case Studies in Chemical and Environmental Engineering Technological Challenges in the Environmental Applications of MXenes and Future Outlook. *Case Stud. Chem. Environ. Eng.* **2**, 100034 (2020).

4

BIODEGRADABLE THERMOPLASTIC POLYMER NANOCOMPOSITES FOR SCREENING ELECTROMAGNETIC RADIATION

Gopika G. Nair and Preema C. Thomas

⇒ **CONTENTS**

4.1 Introduction 35
4.2 Processing Techniques for Biodegradable Thermoplastics 38
4.3 Biodegradable Thermoplastics for EMI Shielding Applications 39
4.4 Poly(Lactic Acid)-Based Nanocomposites for EMI Shielding 40
4.5 Poly(ε-Caprolactone)-Based Nanocomposites for EMI Shielding 44
4.6 Conclusions and Future Perspectives 48
References 50

4.1 INTRODUCTION

As a part of rapid technological growth, each and every domain of society is controlled by electronic systems. Due to the excessive increase in the usage of electronic and communication technologies, we are facing a new problem: electromagnetic pollution largely due to electromagnetic interference (EMI). The misdirected EM radiation emanating from various electronic circuits affects the proper functioning of these devices. Moreover, there are health hazards to humans associated with long-term exposure to EM radiation: it can cause severe damage to human cells, resulting in neurological

DOI: 10.1201/9781003217312-4

disorders.[1] This form of pollution creates chaos in the military, defense, and medical instruments. Thus, the development of proper shielding materials is an area of huge interest.[1]

The very first shielding material was based on metals. Metals such as iron, copper, steel, and aluminum were major shielding materials with good shielding efficiency (SE), because they have good electrical conductivity and appreciable permeability. However, the heavy weight and lack of flexibility retarded the growth of metallic EMI shields. Their poor corrosion resistance was also a major drawback, due to which the use of metal shields is now obsolete.[2–5]

The research to find materials that can satisfy the criteria for promising EMI shielding applications flourished with the development of conducting polymer nanocomposites. These materials could bring about a huge revolution in the field of EMI shielding. Their outstanding processability, electrical conductivity, better flexibility, and corrosion resistance make these nanocomposites better candidates in the field of EMI shields. Moreover, intrinsically conducting polymers bypass metallic EMI shields with inherent properties such as their light weight, scalability, and shielding by absorption mechanism.[6–11] However, in the case of polymer-based shielding materials, the electrical and magnetic properties of the nanocomposites can be effortlessly altered or improved with the incorporation of conducting or magnetic fillers.[1]

When designing EMI shields, some major factors should be considered that are of the utmost importance. The foundation of EMI shielding is based on three mechanisms: shielding by reflection (SE_R), absorption (SE_A), and multiple internal reflections (SE_M), which are represented in Figure 4.1.

Materials with good electrical conductivity have a large distribution of free charge carriers (electrons/holes). Such materials can shield the incoming EM radiation through the interaction of the incident EM field with the free charge carriers. Reflection is the primary shielding mechanism in highly conducting materials like metals.[12,13] In dielectric or magnetic materials, the electric or magnetic dipoles present in the material will interact with the incident EM radiation, and they will absorb energy from it due to the vibration of dipoles. This energy will be lost within the material in the form of heat and, hence, will not be transmitted. This process of shielding by absorption is considered to be better than shielding by reflection because it does not create secondary pollution by back reflection. Materials with good magnetic permeability are also capable of exhibiting the same mechanism of

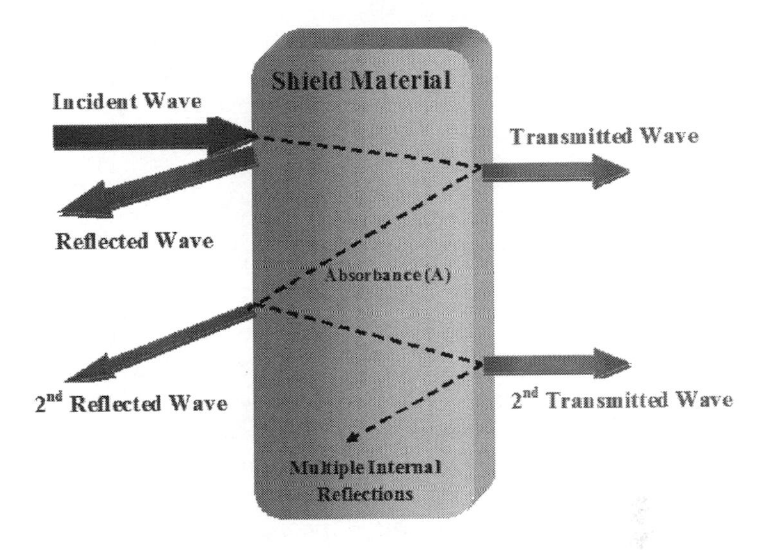

Figure 4.1 EMI shielding mechanism.

shielding.[14,15] Multiple internal reflections are possible for those materials that have a porous structure like aerogels. The three mechanisms of shielding exist in all materials simultaneously at different scales, and the total SE of the material will be the summation of the SE of the three components.[15] The total shielding efficiency of a material can be expressed as follows:

$$SE_T = SE_R + SE_A + SE_M \tag{4.1}$$

As per the requirements for effective shielding, a huge variety of materials can be designed and fabricated. Polymer composites with magnetic fillers are one of the best choices for this purpose. Different structural architectures such as multilayered structures, core-shell structures, and aerogels are also attracting much interest.[16]

The next generation of shielding materials demands an additional criterion to be satisfied. Each and every region of the world is filled with massive amounts of plastic electronic waste material from computers, mobile phones, laptops, and so on. The next revolution is envisaged to be the development of sustainable and biodegradable EMI shielding materials. Recently, many studies have reported on the development of plant-derived, nanocellulose-based EMI shielding materials.[17–20] Materials that are derived from biomaterials can be degraded into harmless biomaterials and are envisaged to overtake plastic-based shielding materials. Biodegradable shields from PLA and PCL can promote the development of technology without sacrificing the sustainability of our environment.[21,22]

4.2 PROCESSING TECHNIQUES FOR BIODEGRADABLE THERMOPLASTICS

Different processing techniques can be employed for the fabrication of biodegradable thermoplastics. Polymer processing techniques can be categorized as molding, extrusion, blow molding, thermoforming, rotational molding, and composite fabrication.[23,24] Molding is the leading plastic processing technique, in which identical parts are manufactured by shaping the plastic material using a common pattern. During the processing of thermoplastics, the polymer in pellet form is introduced into the instrument and then heated above its melting point. As a result, the polymer melts and flows easily into the mold. Injection molding is the fastest molding process. Other processes in the molding category include compression molding, sintering, and transfer molding.[23,24]

Another major technique used for processing plastics is extrusion. It enables the conversion of thermoplastics into products with a uniform shape and density under desired conditions. Extrusion may be divided into two categories: single-screw or twin-screw extrusion. Materials that exhibit constant profile, like window frames, are produced by single-screw extrusion, while twin-screw extrusion is employed to obtain superior dispersion of additives into polymer matrices.[23,24]

Blow molding is another technique employed in the production of hollow articles, like containers, bottles, and jugs. Here, air pressure is employed to force the plastic to melt against the walls of the mold and attain its shape. In the case of the thermoforming technique, plastics in the form of sheets are used to fabricate desired shapes. Initially, the plastic sheets are made soft by heating them to a temperature that is lower than their melting point. Then these molten sheets are pressed around a mold to obtain the shape. Another processing technique is rotational molding or rotomolding, which allows for the fabrication of very large parts and fully encapsulated hollow parts. Polymer in powder form is used for this method, and the enclosed molds are rotated on two axes. The polymer powder within the mold is heated and coats the walls of the mold as it rotates.[23,24] Several other green processing techniques such as 3D printing and fused deposition modeling, along with the use of green solvents such as ionic liquids and lignocellulosic solvents, will also be highly desirable for the fabrication of sustainable, green composites in the near future.[25]

4.3 BIODEGRADABLE THERMOPLASTICS FOR EMI SHIELDING APPLICATIONS

Contemporary society is battling with pollution in various forms in air, water and on land. As a part of the outrageous growth in the electronic and telecommunication sectors, EM pollution is becoming a serious concern. Even though technology is capable of averting the effects of EM pollution predominantly by using plastic-based shields, the aftermath is large heaps of electronic waste at the end of their service life. This non-biodegradable electronic waste (e-waste) is simply being disposed into oceans or landfills, which again gives rise to serious environmental issues.

A sustainable solution to alleviate this issue is to design and fabricate biodegradable EMI shielding materials. Very recently, biodegradable thermoplastics have begun being extensively explored for this application. By the activity of living organisms such as microbes, these biopolymers are easily degradable into less toxic substances and carbon dioxide.[26] Poly(lactic acid) (PLA), poly(Ɛ-caprolactone) (PCL), and chitin are the primary members of the family of biodegradable polymers.[22]

The fundamental property that provides the degradability to most of these polymers is the presence of ester linkages. These linkages facilitate the easy attack of enzymes on these materials, initiating the degradation.[22] There are different classes of biodegradable polymers, which are categorized on the basis of the way in which they are obtained. Polymers like proteins and lignin can be derived from agro sources, whereas poly(hydroxyalkanoate) is a microorganism-derived polymer. Another major class is the biotechnologically derived polymers such as polylactide (PLA), which is synthesized from bioderived monomers. The next class includes polymers like poly(Ɛ-caprolactone) (PCL), which is obtained from petrochemical sources. Most of the biodegradable polymers do not possess sufficient electrical conductivity to be used for EMI shielding applications. But by incorporating suitable nanofillers, one can modify their conductivity, mechanical strength, and magnetic properties, and hence they can be used for the fabrication of shielding materials.[22] By replacing normal plastics or polymers with biopolymers like PLA and PCL, the threat of environmental pollution can be resolved to a large extent.

4.4 POLY(LACTIC ACID)-BASED NANOCOMPOSITES FOR EMI SHIELDING

The world's most widely used bioplastic, poly(lactic acid) (PLA), is a linear aliphatic polyester, which can be derived from biological and renewable materials such as cassava root, cornstarch, sugarcane, and potato.[22,27] The properties of PLA are strongly dependent on various factors, like the type of isomer, the molecular weight of the polymer, and the temperature of the polymerization reaction.[3] It can be synthesized using three methods. The first process is through the formation of lactide followed by ring-opening polymerization. The second one is direct condensation polymerization, and the third is azeotropic dehydration condensation. For the large-scale production of PLA, the technique used is ring-opening polymerization. In the ring-opening polymerization process, PLA is obtained from lactic acid.[28,29]

Lactic acid exists in two stereochemical configurations, namely, L-lactic acid and D-lactic acid, as depicted in Figure 4.2.[28] Initially, this lactic acid is converted into LL-lactide, LD-lactide, or DD-lactide by a catalytic reaction under low pressure. Then it is subjected to ring-opening polymerization, and different forms of PLA, that is, poly(L-lactic acid), poly(D-lactic acid), or poly(LD-lactic acid), are obtained. In the direct condensation and azeotropic dehydration condensation methods, PLA can be obtained directly from L-lactic acid and D-lactic acid.[29]

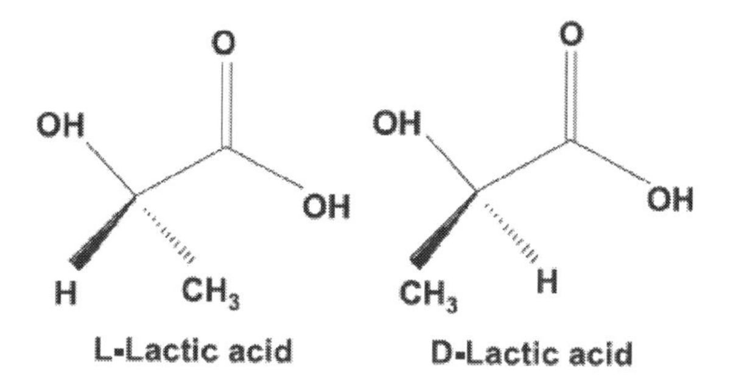

Figure 4.2 Optical isomers of lactic acid.

PLA exists in both semicrystalline (up to 40% crystalline) and amorphous forms. The semicrystalline PLA exhibits better physical and structural properties, which can be attributed to its complex crystalline structure and distribution in the amorphous matrix.[29] PLA has a good aroma barrier; hence, it can be used as packaging material for highly flavored products. It also shows good stability against water, sulfur compounds, aromatic hydrocarbons, ketones, and esters.[29] The surface energy value of PLA indicates that it has a hydrophobic nature compared to other biopolymers.[30] It can be dissolved in solvents like chloroform or chlorinated or fluorinated organic compounds, but it is insoluble in water, alcohols, and alkanes.[30]

This biopolymer can be degraded by both abiotic and biotic degradation. The two main phenomena of abiotic degradation are thermal degradation and hydrolytic degradation. The thermal decomposition of PLA takes places in the temperature range 230–260 °C. Most of the biodegradation occurs as a result of hydrolytic splitting of the ester bonds.[31] The biotic degradation occurs by enzymatic reaction of the polymer.[30]

The possibility of controlling different parameters and properties of PLA creates a wide area, where it can be extremely applicable and beneficial. It has a very low moisture absorption level and low flammability, which make it suitable for sports apparel and products. This polymer is widely used in the manufacture of outdoor furniture due to its high efficiency in resisting UV radiation. Because different structures of PLA can be developed by various spinning techniques and mechanical processing, its application domain ranges from basic apparel to biomedical applications and artificial organs. Because it degrades to biocompatible lactic acid, it is one of the best candidates for medical applications. It is a multidimensional material in the field of medical science.[28] Recently, PLA has been utilized extensively as the matrix material for developing biodegradable EMI shielding materials.

Carbon-based conducting fillers can be easily introduced into a PLA matrix, and composites with moderate conductivity can be designed. Semitransparent and lightweight electromagnetic interference shields were fabricated by Kambiz Chizari et al. via three-dimensional (3D) printing of conductive microstructures using a highly conducting ink made from carbon nanotube (CNT)/PLA nanocomposites.[32] The conducting ink was prepared by mixing the solution of PLA and carbon nanotubes using the ball mill mixing method. Initially, PLA was dissolved in dichloromethane (DCM), whose high volatility makes it a suitable solvent for solution casting, 3D printing (SC3DP) ink. A syringe was filled with CNT/PLA-DCM ink, and the desired pattern was printed with a printing robot controlled by software. SC3DP enables the printing of scaffold structures with different numbers of layers, different interfilament spacing (IFS), and various printing patterns.[32]

Mainly three different scaffold structures were printed namely open window, closed window and zigzag structure which are depicted in Figure 4.2.[32] With increasing CNT, the EMI SE of the PLA scaffold was found to increase due to the presence of more free charge carriers. Also, the transparency can be altered by varying IFS and the number of layers. A maximum SE of 70 dB and transparency of 75% were achieved for a CNT/PLA-DCM conducting ink–printed scaffold.[32]

Magnetic fillers are the best choice to improve magnetic permeability, thereby enhancing EMI shielding via the absorption mechanism. An environmentally friendly, microwave-absorbing material with the desired value of magnetic permeability can be achieved by incorporating magnetic nanoparticles (NPs) into the biopolymer matrix. Moreover, the combination of both dielectric and magnetic materials is an intriguing strategy to obtain high microwave absorption. Quingwei Lu et al. designed a biodegradable EMI shield with effective microwave absorption by embedding Fe_2O_3-anchored, N-doped, carbon nanoparticles in a PLA matrix.[33]

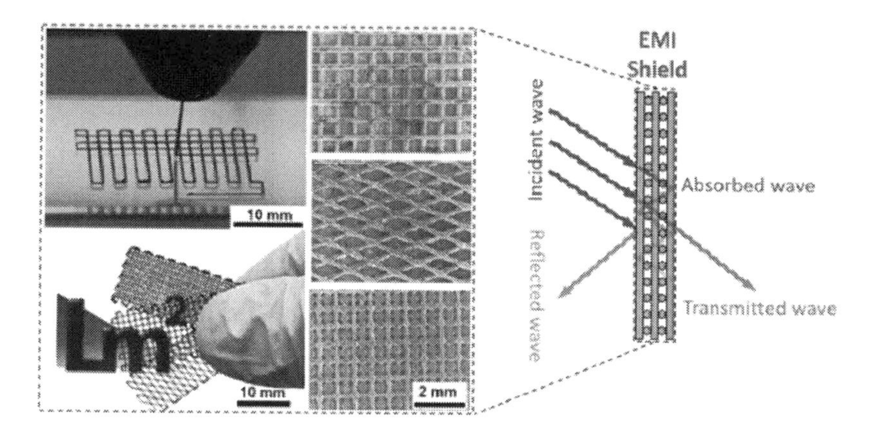

Figure 4.3 3D printing using SC3DP and various microstructures.

In this work, the source carbon was polydopamine (PDA), which was coated onto the Fe_2O_3 particles by the self-polymerization of dopamine (DA), and further proceeded with calcination at 600°C to obtain Fe_2O_3-anchored, N-doped carbon as shown in Figure 4.4a. The TEM images in Figure 4.4b clearly depict the formation of amorphous carbon evenly as multiple layers on the magnetic NPs. The microwave absorption in these hetero-structured nanocomposites was ascribed for a combined effect due to the presence of Fe, N, and carbon. Here, the multiple components with different electronegativity give rise to dipole polarization, and the hetero structure of the sample induces interfacial polarization. As a result, the material absorbs energy through dielectric loss. Magnetic loss of the sample is attributed to the natural resonance and exchange resonance introduced within the material due to the presence of magnetic components. Carbon and iron particles provide free electrons, which form a conductive network in the shielding material that contributes to conduction loss, as depicted clearly in Figure 4.4c.[33] A maximum reflection loss of –60 dB was recorded at a thickness of 2.1 mm. Nevertheless, the absorber immersed in sea water takes only 6 months to degrade and reach a weight loss of 49.35%, and the PLA matrix completely degrades within 1 month. This study successfully demonstrates a strategy for fabricating a good, microwave-absorbing material that is best suited for developing sustainable solutions to EMI shielding.[33]

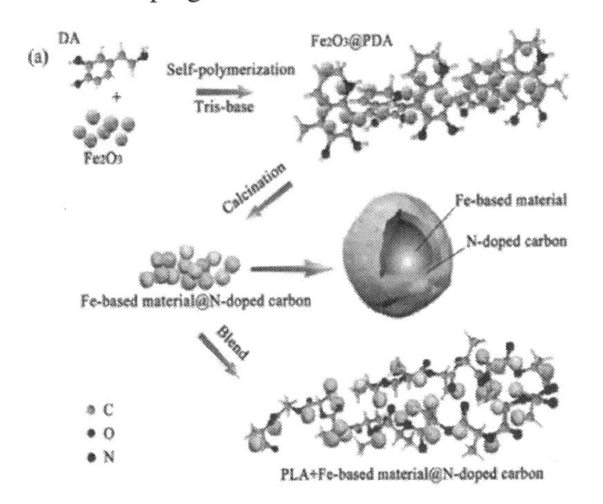

Figure 4.4 (a) Preparation procedure of Fe-based material anchored, N-doped carbon/PLA; (b) TEM images; (c) shielding mechanism of the composite.

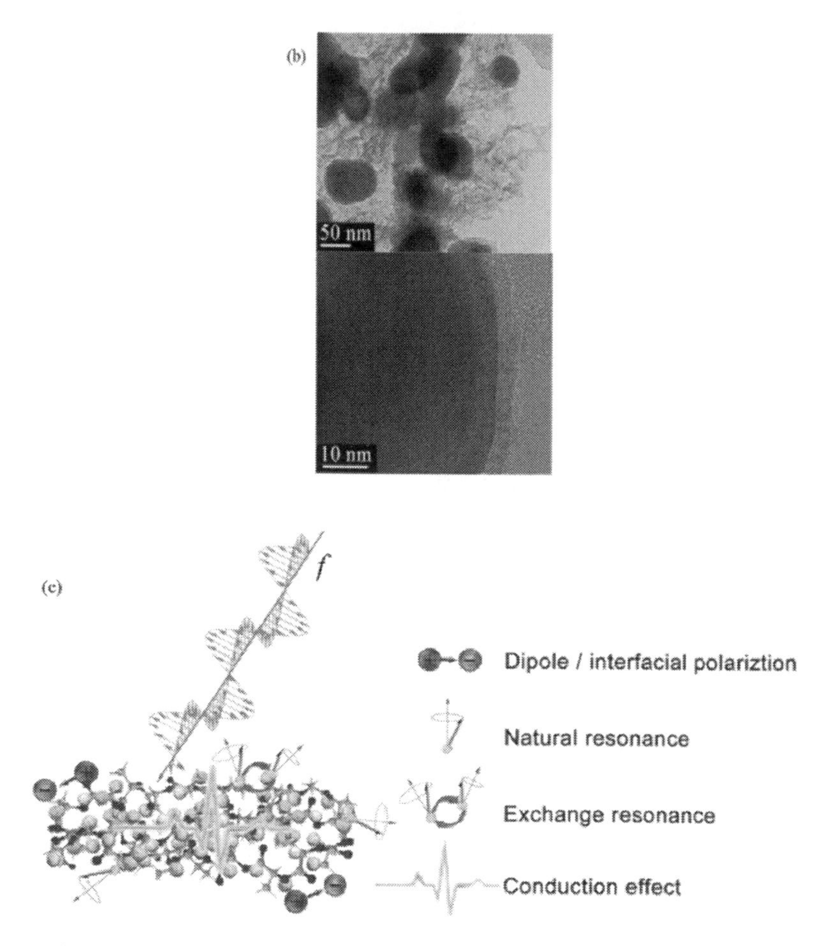

Figure 4.4 (Continued)

4.5 POLY(ε-CAPROLACTONE)-BASED NANOCOMPOSITES FOR EMI SHIELDING

The slow-degrading aliphatic polyester poly(ε-caprolactone) (PCL) is one of the synthetic biopolymers that has been most widely explored for the development of biocompatible scaffolds for biomedical applications.[34,35] Generally, two methods are employed for the synthesis of PCL. The first method is by catalytic ring-opening polymerization of the cyclic ester ε-caprolactone by metal alkoxides, ionic initiators, or metal carboxylates as illustrated in Figure 4.5.[36–39]

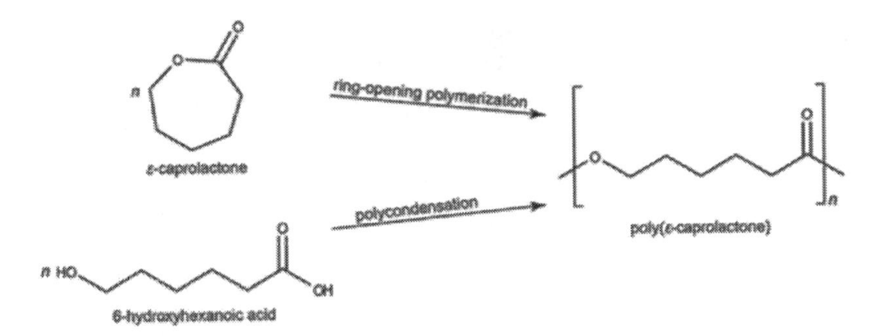

Figure 4.5 Synthesis methods of poly(ε-caprolactone).

The alternate method is the polycondensation of 6-hydroxy-hexanoic acid. This method is less preferred due to the difficulty in attaining a high degree of polymerization. Also, the better quality of the resultant polymer cannot be ensured with this method.[35] Ring-opening polymerization also faces some challenge because this reaction requires a catalyst. Because most of the metal catalysts are toxic and difficult to remove during the purification process, less toxic magnesium- or calcium-based catalysts are preferred, which also possess good activity and low dispersity.[35]

PCL is a semicrystalline polymer with up to 80% crystallinity, which can vary in accordance with variations in cooling rate, trace impurities in the matrix, and molecular weight.[35,40] Toluene, chloroform, and benzene are good solvents for dissolving this biopolymer at room temperature, while it is only sparingly soluble in acetone, acetonitrile, water, and alcohol.[41] The mechanical properties of PCL mostly depend on its crystallinity and molecular weight, and these are also the deciding factors for its use in desired applications.[34,35]

The hydrolytic degradation pathway of PCL can occur by a chemical or enzymatic route. The nonenzymatic random chain splitting by either acid- or base-catalyzed ester hydrolysis is the dominant mechanism of degradation of this synthetic biopolymer. The rate of hydrolytic degradation is found to be highly dependent on the end groups of the polymer chain.[42]

PCL and its composites exhibit outstanding performance as the major materials for biomedical applications. Their biocompatibility, flexibility, and thermoplasticity improve their commercial potential.[43] When compared to PLA, PCL is more stable and takes 2–3 years to completely degrade in biological media. This is very beneficial for the design of scaffolds of tissue engineering applications, because the degradation time is comparable to the regeneration period of human tissue.[44] The biocompatibility and hydrophobic nature of this outstanding polymer make it a superior carrier of lipophilic, hydrophobic, and hydrophilic drugs, and it ensures the homogeneous distribution of drugs at the targeted locations.[45] The mechanical strength, low viscosity, and ease of processing have recently focused attention on PCL as a matrix for EMI shielding polymer composites.[46]

Melt blending is a successful process for the addition of fillers into the matrix of PCL. This is achieved due to the low melting point of PCL. Carbon-based materials are among the best materials to enhance the EMI SE of polymer composites because of their high electrical conductivity. The sandwiched composite of multiwall carbon nanotubes (MWCNTs) and PCL fabricated by melt blending and compression molding, as depicted in Figure 4.6, provides excellent structural architecture with outstanding EMI SE.[46]

In this composite, the PCL layer is insulating and the PCL/MWCNT layer (PCLNT) is electrically conducting. EMI shielding analysis of various combinations of the layers, as shown in Figure 4.7, is conducted for the detailed assessment.[46]

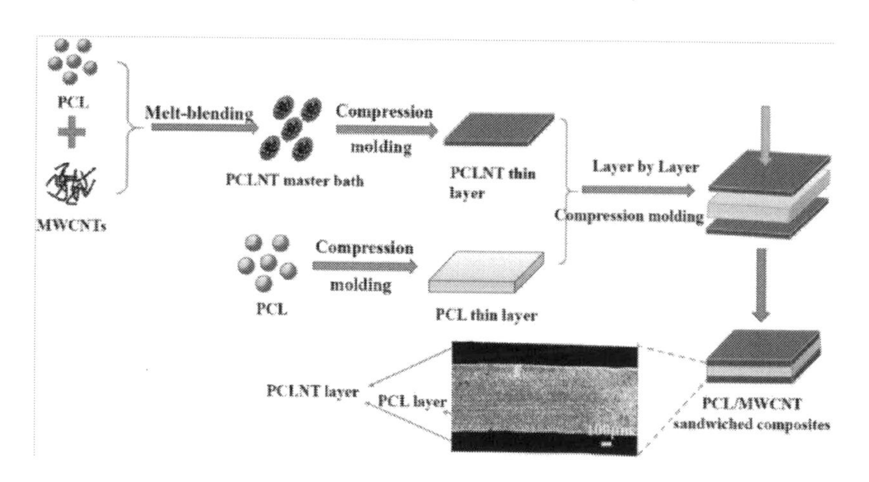

Figure 4.6 Fabrication of the PCL/MWCNT composite.

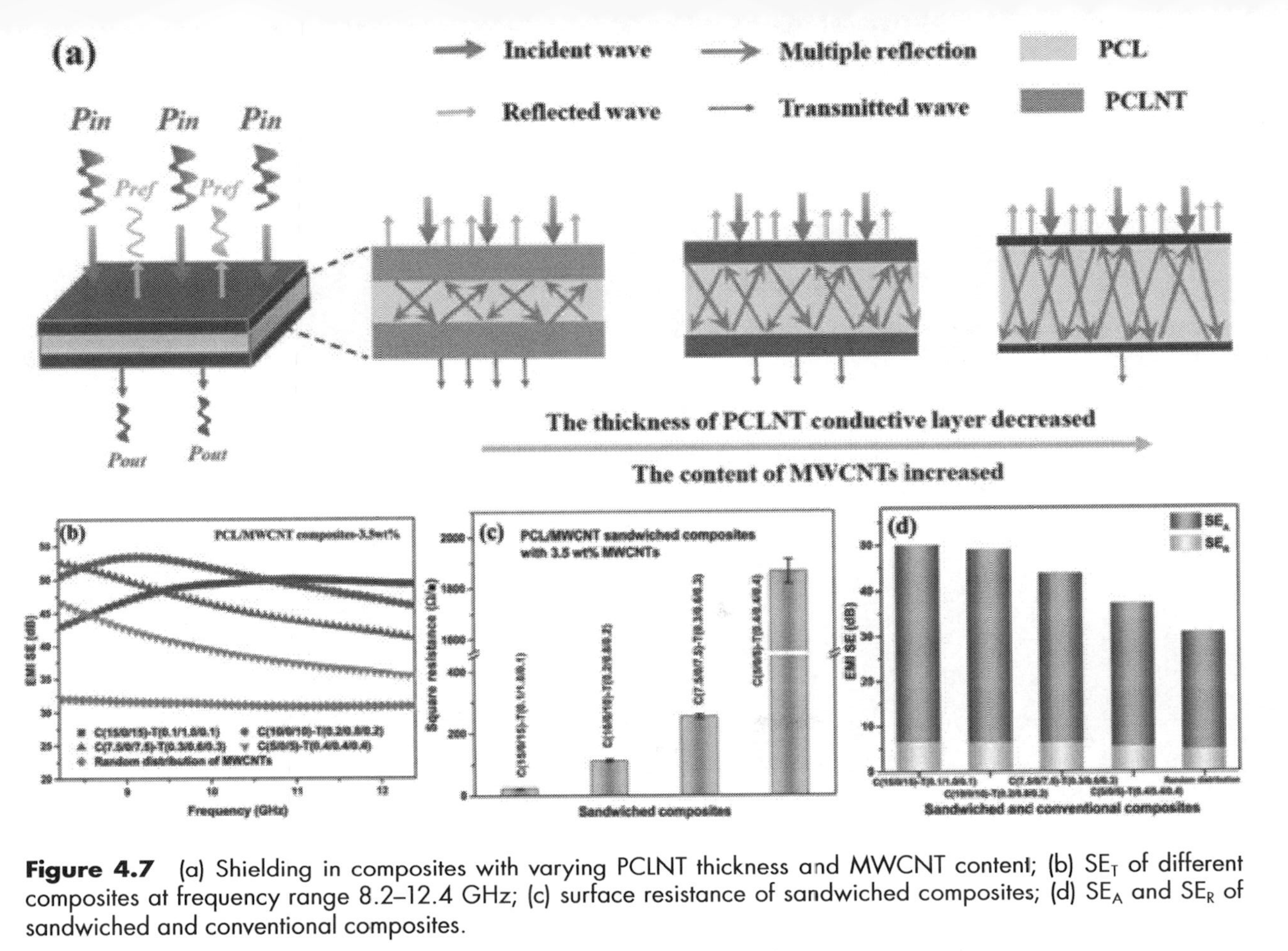

Figure 4.7 (a) Shielding in composites with varying PCLNT thickness and MWCNT content; (b) SE_T of different composites at frequency range 8.2–12.4 GHz; (c) surface resistance of sandwiched composites; (d) SE_A and SE_R of sandwiched and conventional composites.

In this multilayered hierarchical structure, the interface between PCL and PCLNT layers introduces multiple interfaces for internal reflections, which in turn enhances the absorption of the incident radiation. Further modifications in the SE of the multilayered composite were made by increasing the thickness of the conducting PCLNT layer and raising the MWCNT content. As the thickness increases, the square resistivity of the sample decreases, which results in the increases in conductivity and EMI SE value.[46]

Intrinsically conducting polymers like polyaniline (PANI), polypyrrole, and poly(thiophene) are also widely used as fillers in PCL matrices to enhance the overall electrical conductivity of bio-composites. Deepalakshmi et al. studied a combination of reduced graphene oxide (rGO) and PANI in a PCL matrix to develop a flexible, eco-friendly EMI shielding material. The solid films of the composite were synthesized by the solution casting method. The PCL/PANI/rGO hybrid nanocomposite showcased an outstanding SE of 42 dB at 13 GHz, which demonstrates promising potential for these biobased EMI shields for commercial applications in the near future.[47]

4.6 CONCLUSIONS AND FUTURE PERSPECTIVES

The technological advancements in the electronics sector have introduced the challenging scenario of EM pollution. In the past decade, the development of plastic-based EMI shielding materials was considered to be of high importance, and they are now causing serious secondary pollution as electronic waste. According to a recent study, e-waste generation has exceeded 54 million metric tons across the world.[48] It is hoped that the next generation EMI shields will be based on biodegradable polymers and sustainable materials that do not pollute our environment. Biodegradable thermoplastics like PCL and PLA are considered to be promising thermoplastics for the design of biodegradable EMI shields. Some recent studies of these biopolymers are listed in Table 4.1. The future of sustainable EMI shields is based on these biodegradable materials and tuning of the permittivity and permeability of their nanocomposites with suitable nano inclusions.

TABLE 4.1
Comparison of EMI Shielding Performance of Various PLA and PCL Composites Reported in the Literature

Sr No.	Polymer Composite (Matrix/Filler)	Thickness (mm)	Frequency (GHz)	SE_T (dB)	Reference
1	PLA/GNP	1	15	–13	49
2	PLA/Ag nanoplates	2.7	8.2–12.4	–60.4	50
3	PLA/Biochar/Graphite	0.25	18–26.5	–30	51
4	PLA/Graphene Nanoplatelets	1.5	9	–14.3	52
5	PLA/PEG/MWCNT	3	12	–42.07	53
6	PP/PLA/MWCNT	2	8.2–12 4	–13	54
7	PP/PLA/rGO	6	8–12	–22.5	55
8	PLA/Graphite	2	8–12	–45	56
9	PLA/Ag	1.5	8.2–12.4	–50	57
10	PCL/MWCNT	6	18	–32	58
11	PCL-PANI/RGO	5.6	13	–42	47
12	PCL/MWCNT Foam	2 cm	40 MHz to 40 GHz	–80	59
13	PCL/NiO	6	8–12	–12.43	60
14	PCL/Carbonyl Iron Particles	4.83	0.5–20	–34(RL)	61

REFERENCES

1. Kumar P, Narayan Maiti U, Sikdar A, Kumar Das T, Kumar A, Sudarsan V. Recent advances in polymer and polymer composites for electromagnetic interference shielding: Review and future prospects. *Polym Rev.* 2019;59(4):687–738. doi:10.1080/15583724.2019.1625058.
2. Yang Y, Gupta MC, Dudley KL. Studies on electromagnetic interference shielding characteristics of metal nanoparticle- and carbon nanostructure-filled polymer composites in the Ku-band frequency. 2007:85–89. doi:10.1049/mnl:20070042.
3. Yu Y, Ma CM, Teng C, Huang Y, Lee S, Wang I. Electrical, morphological, and electromagnetic interference shielding properties of silver nanowires and nanoparticles conductive composites. *Mater Chem Phys.* 2012;136(2–3):334–340. doi:10.1016/j.matchemphys.2012.05.024.
4. Wenderoth K, Krieger W. Synergism on electromagnetic inductance (EMI)-shielding in metal- and ferroelectric-particle filled polymers. *Polym Compos.* 1989;10(1).
5. Links DA. *Nanoscale.* 2011;1(C):2862–2864. doi:10.1039/c1nr10274e.
6. Wang Y, Jing X. Intrinsically conducting polymers for electromagnetic interference shielding. *Polym Adv Technol.* 2005;16(4):344–351. doi:10.1002/pat.589.
7. Kessler Z. I I : : 14(2).
8. Mbkelga T, Jussilaa SPTTS, Isotaloa H. Thin polyaniline films in EMI shielding. *Synth Met.* 1997;85:1335–1336.
9. Courric S, Tran VH. The electromagnetic properties of poly (p-phenylene-vinylene) derivatives. *Polymer.* 1998;39(12):2399–2408.
10. Cable LA. Industrial applications of polyaniline. *Synth Met.* 1999;101:703–704.
11. Pardoen T, Bailly C, Thomassin J, Je C, Huynen I, Detrembleur C. Polymer/carbon based composites as electromagnetic interference (EMI) shielding materials. Published online 2013. doi:10.1016/j.mser.2013.06.001.
12. Lai K, Sun RJ, Chen MY, Wu H, Zha AX. Electromagnetic shielding effectiveness of fabrics with metallized polyester filaments. *Text Res J.* 2007;77(4):242–246. doi:10.1177/0040517507074033.
13. Roh JS, Chi YS, Kang TJ, Nam SW. Electromagnetic shielding effectiveness of multifunctional metal composite fabrics. *Text Res J.* 2008;78(9):825–835. doi:10.1177/0040517507089748.
14. Abdi MM, Kassim AB, Ekramul Mahmud HNM, Yunus WMM, Talib ZA. Electromagnetic interference shielding effectiveness of new conducting polymer composite. *J Macromol Sci Part A Pure Appl Chem.* 2010;47(1):71–75. doi:10.1080/10601320903399834.

15. Sambyal P, Singh AP, Verma M, Farukh M, Singh BP, Dhawan SK. Tailored polyaniline/barium strontium titanate/expanded graphite multiphase composite for efficient radar absorption. *RSC Adv.* 2014;4(24):12614–12624. doi:10.1039/c3ra46479b.
16. Wilson R, George G, Joseph K. *An Introduction to Materials for Potential EMI Shielding Applications: Status and Future.* Elsevier Inc.; 2020. doi:10.1016/b978-0-12-817590-3.00001-4.
17. Pai AR, Paoloni C, Thomas S. Nanocellulose-based sustainable microwave absorbers to stifle electromagnetic pollution. In: *Nanocellulose Based Composites for Electronics.*; 2021:237–258. doi:10.1016/b978-0-12-822350-5.00010-2.
18. Gopakumar DA, Pai AR, Pottathara YB, et al. Cellulose nanofiber-based polyaniline flexible papers as sustainable microwave absorbers in the X-Band. *ACS Appl Mater Interfaces.* 2018;10(23):20032–20043. doi:10.1021/acsami.8b04549.
19. Pai AR, Binumol T, Gopakumar DA, et al. Ultra-fast heat dissipating aerogels derived from polyaniline anchored cellulose nanofibers as sustainable microwave absorbers. *Carbohydr Polym.* 2020;246:116663. doi:10.1016/j.carbpol.2020.116663.
20. Gopakumar DA, Pai AR, Pottathara YB, et al. Flexible papers derived from polypyrrole deposited cellulose nanofibers for enhanced electromagnetic interference shielding in gigahertz frequencies. *J Appl Polym Sci.* 2021;138(16):1–11. doi:10.1002/app.50262.
21. Kargarzadeh H, Huang J, Lin N, et al. Recent developments in nanocellulose-based biodegradable polymers, thermoplastic polymers, and porous nanocomposites. *Prog Polym Sci.* 2018;87(August):197–227. doi:10.1016/j.progpolymsci.2018.07.008.
22. Nath K, Bhattacharyya SK, Das NC. *Biodegradable Polymeric Materials for EMI Shielding.* Elsevier Inc.; 2020. doi:10.1016/b978-0-12-817590-3.00010-5.
23. Cantor KM, Watts P. *Plastics Processing.* Elsevier; 2011. doi:10.1016/B978-1-4377-3514-7.10012-1.
24. Polymers L, Plastics T, Polymers I, et al. *Plastics Processing 1.* Kirk-Othmer Encyclopedia of Chemical Technology.
25. Joseph B, Krishnan S, Kavil SV, et al. Green chemistry approach for fabrication of polymer composites. *Sustain Chem.* 2021;2(2): 254–270. doi:10.3390/suschem2020015.
26. Ammala A, Bateman S, Dean K, et al. *An Overview of Degradable and Biodegradable Polyolefins.* Vol. 36. Elsevier Ltd; 2011. doi:10.1016/j.progpolymsci.2010.12.002.
27. Sha L, Chen Z, Chen Z, Zhang A, Yang Z. Polylactic acid based nanocomposites: Promising safe and biodegradable materials in biomedical field. *Int J Polym Sci.* 2016;2016. doi:10.1155/2016/6869154.
28. Gupta B, Revagade N. Poly (lactic acid) fiber : An overview. 2007;32:455–482. doi:10.1016/j.progpolymsci.2007.01.005.

29. POLY (LACTIC ACID). Published online 2010.
30. Belgacem MN, Gandini A. Monomers, Polymers and Composites from Renewable Resources. In: *Comprehensive Polymer Science* (eds. S.L. Aggarwal and S. Russo), 1992, 528–573. Oxford: Pergamon Press.
31. Li SM, Garreau H, Vert M. Structure-property relationships in the case of the degradation of massive poly(α-hydroxy acids) in aqueous media—Part 2 Degradation of lactide-glycolide copolymers: PLA37.5GA25 and PLA75GA25. *J Mater Sci Mater Med.* 1990;1(3):131–139. doi:10.1007/BF00700872.
32. Chizari K, Arjmand M, Liu Z, Sundararaj U, Therriault D. Three-dimensional printing of highly conductive polymer nanocomposites for EMI shielding applications. *Mater Today Commun.* 2017;11:112–118. doi:10.1016/j.mtcomm.2017.02.006.
33. Liu Q, Zeng M, Liu J, et al. Fe-based material@N-doped carbon composites as environment-friendly microwave absorbers. *Carbon N Y.* 2021;171(November 2020):646–657. doi:10.1016/j.carbon.2020.09.045.
34. Bartnikowski M, Dargaville TR, Ivanovski S, Hutmacher DW. Degradation mechanisms of polycaprolactone in the context of chemistry, geometry and environment. *Prog Polym Sci.* 2019;96: 1–20. doi:10.1016/j.progpolymsci.2019.05.004.
35. Labet M, Thielemans W. Synthesis of polycaprolactone: A review. *Chem Soc Rev.* 2009;38(12):3484–3504. doi:10.1039/b820162p.
36. Malinová L, Brožek J. Ethyl magnesium bromide as an efficient anionic initiator for controlled polymerization of ε-caprolactone. *Polym Bull.* 2014;71(1):111–123. doi:10.1007/s00289-013-1048-3.
37. Limwanich W, Meepowpan P, Nalampang K, Kungwan N, Molloy R, Punyodom W. Kinetics and thermodynamics analysis for ring-opening polymerization of ε-caprolactone initiated by tributyltin n-butoxide using differential scanning calorimetry. *J Therm Anal Calorim.* 2015;119(1):567–579. doi:10.1007/s10973-014-4111-x.
38. Berl M, Kricheldorf HR, Scharnagl N. Polymerisation mechanism of metal alkoxide initiated polymerisations of lactide and various lactones. *Macromolecules.* 1988;21(12):286.
39. Agarwal S. Chemistry, chances and limitations of the radical ring-opening polymerization of cyclic ketene acetals for the synthesis of degradable polyesters. *Polym Chem.* 2010;1(7):953–964. doi:10.1039/c0py00040j.
40. Pitt CG, Chasalow FI, Hibionada YM, Klimas DM, Schindler A. Aliphatic polyesters. I. The degradation of poly(ϵ-caprolactone) in vivo. *J Appl Polym Sci.* 1981;26(11):3779–3787. doi:10.1002/app.1981.070261124.
41. Sinha VR, Bansal K, Kaushik R, Kumria R, Trehan A. Poly-ε-caprolactone microspheres and nanospheres: An overview. *Int J Pharm.* 2004;278(1):1–23. doi:10.1016/j.ijpharm.2004.01.044.

42. Ouhadi T, Stevens C, Teyssié P. Study of poly-ϵ-caprolactone bulk degradation. *J Appl Polym Sci.* 1976;20(11):2963–2970. doi:10.1002/app.1976.070201104.
43. Mohamed RM, Yusoh K. A review on the recent research of polycaprolactone (PCL). *Adv Mater Res.* 2015;1134:249–255. doi:10.4028/www.scientific.net/amr.1134.249.
44. Malikmammadov E, Tanir TE, Kiziltay A, Hasirci V, Hasirci N. *PCL and PCL-Based Materials in Biomedical Applications.* Vol. 29. Taylor & Francis; 2018. doi:10.1080/09205063.2017.1394711.
45. Wang X, Wang Y, Wei K, Zhao N, Zhang S, Chen J. Drug distribution within poly(ε-caprolactone) microspheres and in vitro release. *J Mater Pro.cess Technol.* 2009;209(1):348–354. doi:10.1016/j.jmatprotec.2008.02.004.
46. Tang XH, Li J, Wang Y, Weng YX, Wang M. Controlling distribution of multi-walled carbon nanotube on surface area of Poly(ε-caprolactone) to form sandwiched structure for high-efficiency electromagnetic interference shielding. *Compos Part B Eng.* 2020;196(April):108121. doi:10.1016/j.compositesb.2020.108121.
47. Ponnamma D, Sadasivuni KK, Strankowski M, Kasak P, Krupa I, AlMaadeed MAA. Eco-friendly electromagnetic interference shielding materials from flexible reduced graphene oxide filled polycaprolactone/polyaniline nanocomposites. *Polym—Plast Technol Eng.* 2016;55(9):920–928. doi:10.1080/03602559.2015.1132435.
48. Foreword.:8–9.
49. Kashi S, Gupta RK, Bhattacharya SN, Varley RJ. Experimental and simulation study of effect of thickness on performance of (butylene adipate-co-terephthalate) and poly lactide nanocomposites incorporated with graphene as stand-alone electromagnetic interference shielding and metal-backed microwave abso. *Compos Sci Technol.* 2020;195(April):108186. doi:10.1016/j.compscitech.2020.108186.
50. Li J, Peng WJ, Fu ZJ, et al. Achieving high electrical conductivity and excellent electromagnetic interference shielding in poly(lactic acid)/silver nanocomposites by constructing large-area silver nanoplates in polymer matrix. *Compos Part B Eng.* 2019;171(April): 204–213. doi:10.1016/j.compositesb.2019.05.003.
51. Tolvanen J, Hannu J, Hietala M, Kordas K, Jantunen H. Biodegradable multiphase poly(lactic acid)/biochar/graphite composites for electromagnetic interference shielding. *Compos Sci Technol.* 2019; 181(March):107704. doi:10.1016/j.compscitech.2019.107704.
52. Kashi S, Gupta RK, Baum T, Kao N, Bhattacharya SN. Morphology, electromagnetic properties and electromagnetic interference shielding performance of poly lactide/graphene nanoplatelet nanocomposites. *Mater Des.* 2016;95:119–126. doi:10.1016/j.matdes.2016.01.086.

53. Ahmad AF, Aziz SA, Obaiys SJ, et al. Biodegradable poly (lactic acid)/poly (ethylene glycol) reinforced multi-walled carbon nanotube nanocomposite fabrication, characterization, properties, and applications. *Polymers (Basel)*. 2020;12(2):1–22. doi:10.3390/polym12020427.
54. Soares BG, Cordeiro E, Maia J, Pereira ECL, Silva AA. The effect of the noncovalent functionalization of CNT by ionic liquid on electrical conductivity and electromagnetic interference shielding effectiveness of semi-biodegradable polypropylene/poly(lactic acid) composites. *Polym Compos*. 2020;41(1):82–93. doi:10.1002/pc.25347.
55. Ahmad AF, Ab Aziz S, Abbas Z, et al. Chemically reduced graphene oxide-reinforced poly(lactic acid)/poly(ethylene glycol) nanocomposites: Preparation, characterization, and applications in electromagnetic interference shielding. *Polymers (Basel)*. 2019;11(4). doi:10.3390/polym11040661.
56. Wang G, Zhao G, Wang S, Zhang L, Park CB. Injection-molded microcellular PLA/graphite nanocomposites with dramatically enhanced mechanical and electrical properties for ultra-efficient EMI shielding applications. *J Mater Chem C*. 2018;6(25):6847–6859. doi:10.1039/c8tc01326h.
57. Zhang K, Yu HO, Yu KX, et al. A facile approach to constructing efficiently segregated conductive networks in poly(lactic acid)/silver nanocomposites via silver plating on microfibers for electromagnetic interference shielding. *Compos Sci Technol*. 2018;156:136–143. doi:10.1016/j.compscitech.2017.12.037.
58. Pawar SP, Kumar S, Misra A, Deshmukh S, Chatterjee K, Bose S. Enzymatically degradable EMI shielding materials derived from PCL based nanocomposites. *RSC Adv*. 2015;5(23):17716–17725. doi:10.1039/c4ra10364e.
59. Thomassin JM, Pagnoulle C, Bednarz L, Huynen I, Jerome R, Detrembleur C. Foams of polycaprolactone/MWNT nanocomposites for efficient EMI reduction. *J Mater Chem*. 2008;18(7):792–796. doi:10.1039/b709864b.
60. Ahmad AF, Abbas Z, Aziz SA, Obaiys SJ, Zainuddin MF. Synthesis and characterisation of nickel oxide reinforced with polycaprolactone composite for dielectric applications by controlling nickel oxide as a filler. *Results Phys*. 2018;11:427–435. doi:10.1016/j.rinp.2018.08.041.
61. Rath U, Pandey PM. Investigations into the microwave shielding behavior of oriented Polycaprolactone/Carbonyl iron particles composites fabricated using magnetic field assisted extrusion 3D printing. *Proc Inst Mech Eng Part C J Mech Eng Sci*. Published online 2020. doi:10.1177/0954406220959098.

5

CONDUCTING POLYMER-BASED MATERIALS FOR ELECTROMAGNETIC INTERFERENCE SHIELDING APPLICATIONS

Deepa K. Baby

⇒ **CONTENTS**

5.1 Introduction	55
5.2 EMI Shielding with PANI	57
5.3 EMI Shielding with Polypyrrole (PPY)	58
5.4 EMI Shielding by Electrically Conducting Graphene-Based Polyurethane Nanocomposites	59
5.5 EMI Shielding by Dispersed Metal Nanowires in a Polystyrene Matrix	60
5.6 EMI Shielding by CNT/PLA Conductive Nanocomposite Scaffolds	61
5.7 Conclusion	61
References	62

5.1 INTRODUCTION

The widespread use of electronic, computing, and telecommunication equipment has raised the problem of electromagnetic interference (EMI), also called electromagnetic *pollution*. EMI manifests itself as a disturbance in the operation of electronic devices, which ultimately can lead to improper operation or malfunction of the equipment. Metals are common materials that have been employed as EMI shields; however, metallic shields have drawbacks like corrosion, high cost, high weight, and expensive to process. Hence, during the last decade, technological breakthroughs and research focus

DOI: 10.1201/9781003217312-5

in the field of EMI shielding materials have been intensely directed toward the development of conducting polymer-based nanomaterials (CPNs).[1] Intrinsically conducting polymers are organic polymers that possess the electrical, electronic, magnetic, and optical properties of a metal, while retaining the mechanical properties and processability[2] commonly associated with a conventional polymer, more commonly known as "synthetic metals." Due to their excellent qualities such as low weight, low cost, easy processability, corrosion resistance, and versatility, conductive polymer composites are promising materials for use in enclosures for electrical and electronic instruments to protect them from electromagnetic (EM) waves and also to fulfill electromagnetic interference (EMI) shielding requirements. Electrical conductivity is a key parameter for effective EMI shielding materials. Most common polymers are inherently insulative; however, the embedding of a sufficient amount of conductive nanofiller into a polymer matrix leads to the formation of conductive networks across the nanocomposite.[3] This transforms the whole nanocomposite into a conductive material.[4] Conduction via physical contact between conductive nanofillers in combination with electron tunnelling and hopping between conductive nanofillers is the main mechanism for electron transference in CPNs.

The electrical and electromagnetic properties of polymeric materials filled with conducting nanofillers have been investigated by many researchers.[5] The nanocomposites filled with high aspect ratio conductive nanofillers exhibit higher EMI shielding and a lower electrical percolation threshold than the traditional composites based on microfillers such as carbon fiber. As nanofillers based on carbon nanotubes are much more expensive than polymers, there is still a need for further reduction in nanofiller loading to improve the competitiveness of CNT/polymer nanocomposites.[6]

Electromagnetic energy consists of a magnetic (H) field and an electric (E) field component perpendicular to each other that propagate at right angles to the plane containing the two components.[7] The ratio of E to H is defined as the wave impedance (Z_w, in ohms) and depends on the type of source and the distance from the source. Large impedances characterize electric fields, and small ones characterize magnetic fields. Far from the source, the ratio of E to H remains constant. In EMI shielding, there are two type of regions: the near field shielding region and far field shielding region. When the distance between the radiation source and the shield is larger than $\lambda/2\pi$ (where λ is the wavelength of the source), it is in the far field shielding region. Electromagnetic plane wave theory is generally applied for EMI shielding in this region. When the distance is less than $\lambda/2\pi$, it is in the near field shielding region, and the theory based on the contribution of electric and magnetic dipoles is used for EMI shielding. The amount of attenuation offered by a shield depends on three mechanisms. The first is usually a reflection of the wave from the shield. The second is absorption of the wave into the shield as it passes through it. The third is due to the re-reflections, that is, the multiple reflections of the waves at various surfaces or interfaces in the shield.

Shielding effectiveness (SE)[8] in decibels (dB) is a measure of the reduction in EMI at a specific frequency achieved by a shield, such as a coating, and is defined as follows:

$$SE = 10\,Log\,P_o\,/\,P_t \tag{5.1}$$

where P_0 is the power intensity incident on the shield, and P_t is the counterpart transmitted through the shield.

5.2 EMI SHIELDING WITH PANI

The first reported study of polyaniline (PANI) for EMI shielding was made by Shacklette[9] and co-workers,[10] in which PANI was melt blended with polyvinyl chloride (PVC) to form a composite with a conductivity of 20 S/cm. The EMI shielding effectiveness (SE) of the PANI composites, as well as that of other composites of metal fillers, was measured over a frequency range from 1 MHz to 3 GHz and calculated theoretically for both near and far fields. The measured SE agreed well with the theoretical calculations in both the near and far fields. The higher the conductivity, the higher the SE of the composite. It can be concluded that, as long as the distribution of the conductive fillers is uniform, which is difficult for metal fillers to achieve, and the size of the filler is smaller than the wavelength of the radiation, the frequency dependence of the SE was independent of the nature of the filler.

Conductivity is the only important parameter in calculating the SE of metals due to their higher microwave frequency loss tangent (tan δ), which is far more than 1. However, in the case of intrinsically conductive polymers (ICPs), the permittivity and tan δ are both important for the calculation of their SE due to the lower tan δ of ICPs, which is between 0.5 and 7.0.[11] For a given doped PANI sample, the SE increased linearly with thickness, except in the case of thicknesses below 10 mm where the SE varied nonlinearly due to the multiple reflections inside the PANI film.[12] Stretching of the film also increases the SE value, and temperature has no effect. Intrinsic parameters, such as permittivity (both the real and imaginary parts and the absolute value of complex permittivity) and tan δ, controlled the SE whether due to absorption or reflection.[13] To increase the SE or conductivity of PANI, mixtures of PANI with conducting powders including silver, graphite, and carbon black were prepared. PANI was mixed separately with these powders and then dissolved in *N*-methyl-2-pyrrolidone (NMP). The solutions were cast to form the corresponding films, which were doped with hydrochloric acid (HCl).[14] EMI shielding measurements, which were carried out according to ASTM D4935-89 at room temperature in the frequency range of 10 MHz to 1 GHz, showed that the SE of all films was independent of the testing frequency and was approximately in agreement with the theoretical SE calculated by using plane wave theory.[15]

Free-standing PANI films of different thicknesses, prepared by solution cast and doped with HCl thereafter, were tested for EMI shielding, and EMI SE, measured using both holders, agreed well in the range of 50 MHz to 1.5 GHz, as well as with the theoretical SE calculated based on conductivities. The thicker the films, the higher the SE values. For example, a film of 20 mm has a SE of 6.1 and 4.6 dB with the new holder and ASTM holder, respectively; a film of 90 mm has a SE of 18.6 and 17.6 dB. In the higher frequencies, the reflectance and absorbance for the 90-mm PANI film were ca. 70% and 30%, respectively, suggesting that absorption is the main mechanism for PANI in EMI shielding.[16]

5.3 EMI SHIELDING WITH POLYPYRROLE (PPY)

The conductivities of electropolymerized PPY films increased with the concentration of dopant. The microwave transmission decreased with increased dopant concentration (or increased conductivity); for example, at very low conductivities, the transmission was high with very little reflection. The microwave reflection increased with dopant concentration; for instance, at higher conductivities (10–50 S/cm), microwave reflection was very high with nearly no transmission.[17] For microwave absorption, there was first an increase and then a decrease with the increase in dopant concentration. All of the microwave transmission and reflection results with conductivities and thicknesses were well modeled using a high-frequency structure simulator, which will provide a valuable guide when using the films for microwave applications such as EMI shielding and radar absorption.[18] A significant SE of 38 dB was achieved with highly doped conductive film, while the lightly doped semiconductive film exhibited a small SE value. The far field SE of PPY films increases with temperature. For instance, at room temperature, the film with a conductivity of 0.01 S/cm was transparent to microwaves, and as the temperature increased, the SE increased significantly because of the increase in conductivity.[19] The far field SE of the films decreased with an increase in frequency from 300 MHz to 2 GHz, while the near field SE increased with frequency from 200 MHz to 1 GHz, which contained both electric and magnetic field shielding. PPY-based, intrinsically conducting, hot melt adhesives (ICHMAs) retain the advantages of classical hot melt adhesives while incorporating significant EMI shielding properties. Such ICHMAs are especially appropriate for use in electronic, computing, and telecommunication applications requiring a lightweight, quick-bonding, EMI-shielding adhesive or sealant.[20] Also, they can be employed to cover intricate surfaces, small areas, or elements that irradiate and/or receive electromagnetic energy out of specified values.

5.4 EMI SHIELDING BY ELECTRICALLY CONDUCTING GRAPHENE-BASED POLYURETHANE NANOCOMPOSITES

Thermoplastic polyurethane (TPU) polymer has immense potential applications in the fields of smart actuators, high-performance coatings, and adhesives because of its excellent chemical abrasion, weather resistance, flexibility, and transparency properties.[21] The low tensile strength, poor gas barrier, and electrical properties of TPU applications can be improved by incorporating carbon nanofillers such as graphene into a TPU matrix.[22] The conducting carbon nanofillers act as reinforcing material and also develop an electrical conducting matrix for advanced applications such as EMI shielding.[23] Nanni et al.[24] studied the microwave shielding properties of expanded graphite-based PU composites and observed a shielding effectiveness (SE) of –20 dB in the X-band microwave frequency region of 8.2–12.4 GHz at 20 wt% loading of expanded graphite (EG) at 4-mm thickness. The graphene-like carbon nanostructures (GNCs) incorporated in the PU matrix in the X-band region for 25 wt% loaded GNCs/PU showed 26.45 dB at 2-mm thickness of the sample.[25] The electrical conductivity of the material plays a crucial role in attaining effective EMI shielding. Generally, the shielding efficiency increases with the increasing electrical conductivity of the composite.[26] The electrical conductivity of TPU/TRG nanocomposites increases with an increase in thermally reduced grapheme (TRG) nanosheet loading. The DC conductivity of composites increases from 2.81×10^{-13} S/cm (neat TPU) to 3.1×10^{-4} S/cm with 5.5 vol% loading of TRG. EMI shielding values were found to be dominated by the material's absorption behavior.

5.5 EMI SHIELDING BY DISPERSED METAL NANOWIRES IN A POLYSTYRENE MATRIX

The electrical conductivity of polymer nanocomposites is resolved by the composition of the material (type and concentration of conductive nanoparticles), the dispersion, distribution, alignment, surface properties, and aspect ratio of the nanoparticles, the properties of the polymer matrix, and processing conditions.[27] The final morphology of the composites and the electrical connectivity between conductive nanoparticles are critical for the electrical properties of these advanced and novel materials.[28]

Below the percolation threshold, there is no electrical conductivity, and metal nanowire embedded polystyrene (NEPS) is similar to that of unfilled polymer because the nanowires are not close enough to enable electron transfer.[29] At a critical copper metal nanowire (CuNW) volume fraction between 0.006 and 0.008, there is a sudden and sharp increase of 12 orders of magnitude in electrical conductivity 6, indicating the formation of electrically conductive nanowire networks. An increase in CuNW volume fraction from 0.008 to 0.028 resulted in a gradual increase in 6 from 101 to 104 S/m, respectively. These CuNW/PS nanocomposites exhibit some of the highest electrical conductivities ever reported for conductive polymer nanocomposites with low filler content. Electrically conductive networks of nanowires result in materials with high EMI shielding effectiveness.[30] EMI shielding effectiveness (SE) is a measure of the material's ability to attenuate the intensity of EM waves. EMI SE of CuNW/PS composites increases with increasing CuNW concentration. For instance, EMI SE increased from 6.5 to 38 dB with an increase in CuNW concentration from 0.8 to 1.8 vol%. EMI SE of samples with concentrations higher than 2.0 vol% CuNW and 210 mm in thickness were beyond the dynamic range of the characterization equipment (>50 dB). Nanocomposites with concentrations higher than 1.0 vol%

showed EMI SE > 20 dB and will be suitable for desktop and laptop applications. The EMI SE can be increased by increasing the thickness of the shield. Thus, it is expected that metal nanowire–based polymer nanocomposites with thicknesses greater than 210 mm will exhibit even higher EMI SE and will be suitable for more advanced applications, like medical and military equipment where a very high level of shielding is demanded.

5.6 EMI SHIELDING BY CNT/PLA CONDUCTIVE NANOCOMPOSITE SCAFFOLDS

One of the main advantages of the scaffold structures over the solid EMI shielding structures is the semitransparency of these gridlike structures, which can be useful for applications where transparency is key. The EMI SE of the carbon nanotube/poly lactic acid (CNT/PLA) shields in both types of structures, scaffold and solid forms, increased with increasing CNT concentration.[31] For the solid samples, EMI SE increased to about 47 and 55 dB for CNT concentrations of 20 and 30 wt%, respectively. In order to test the reproducibility of the results, three identical scaffolds with 20 wt% CNT were fabricated and their EMI SE was measured. The EMI SE of the fabricated CNT/PLA in solid form with high CNT loading (≥10 wt%), demonstrates the importance of the thickness of the characterized nanocomposite films. The increase in EMI shielding with CNT content can be attributed to the increase in shielding by both EM reflection and EM absorption.[32] When an EM wave hits a conductive shield, a fraction of the EM wave is reflected off the shield due to interaction with the surface free charge carriers, and another fraction infiltrates through the shield with its energy dissipated via absorption.[33] The materials with higher conductivity present higher reflection, while the magnetic materials reduce shielding by reflection. Shielding by absorption for a conductive material is proportional to electrical conductivity. The materials with high conductivity and high magnetic permeability attenuate EM waves efficiently. Absorption attenuates EM waves through interaction with free charge carriers and electric or magnetic dipoles.

5.7 CONCLUSION

Electroconductive, flexible, water-resistant polymeric conducting materials are very useful for mitigating electromagnetic radiation pollution, which could be their unique selling point as a commercially marketable low-cost product for advanced applications in the fields of space and aviation and in other lightweight devices. EMI SE depends on the thickness of the material. EM attenuation has a direct correlation with variations in thickness. The EMI SE of conducting polymer nanocomposites relies on numerous factors, such as high electric conductivity, a greater number of free mobile charge carriers, conductive mesh formation that decreases with increasing CNF nanofiller loading, interaction of conducting and absorbing sites, proper dispersion and distribution of nanofiller, efficient electron transport, higher absorption that is dependent on dielectric properties including conduction loss, dipolar relaxation, thickness, and electrical polarization.

REFERENCES

1. Thomassin JM, Jérôme C, Pardoen T, Bailly C, Huynen I, Detrembleur C, Polymer/carbon based composites as electromagnetic interference (EMI) shielding materials, *Mater. Sci. Eng. R-Rep.* 74 (2013) 211–232.
2. Arjmand M, Apperley T, Okoniewski M, Sundararaj U, Comparative study of electromagnetic interference shielding properties of injection molded versus compression molded multi-walled carbon nanotube/polystyrene composites, *Carbon* 50 (2012) 5126–5134.
3. Ameli A, Jung P, Park C, Electrical properties and EMI shielding effectiveness of polypropylene/carbon fiber composite foams, *Carbon* 60 (2013) 379–391.
4. Arjmand M, Mahmoodi M, Park S, Sundararaj U, Impact of foaming on the broadband dielectric properties of multi-walled carbon nanotube/polystyrene composites, *J. Cell. Plast.* 50 (2014) 551–562.
5. Alig I, Pötschke P, Lellinger D, Skipa T, Pegel S, Kasaliwal GR, et al., Establishment, morphology and properties of carbon nanotube networks in polymer melts, *Polymer* 53 (January 1, 2012) 4–28.
6. Al-Saleh MH, Sundararaj U, Electrically conductive carbon nanofiber/ polyethylene composite: Effect of melt mixing conditions, *Polym. Adv. Technol.* 22 (2) (2011) 246–253.
7. Chung DDL, Materials for electromagnetic interference shielding, *J. Mater. Eng. Perform.* 9 (2000) 350.
8. Violette JLN, White DRJ, Violette MF, *Electromagnetic Compatibility Handbook*, Van Nostrand Reinhold Company, New York, 1987.

9. Shacklette LW, Colaneri NF, Kulkarni VG, Wessling B, EMI shielding of intrinsically conductive polymers, *J. Vinyl Technol.* 14 (1992) 118.
10. Colaneri NF, Shacklette LW, EMI shielding measurements of conductive polymer blends. *IEEE Trans. Instrum. Meas.* 41 (1992) 291.
11. Epstein AJ, MacDiarmid AG, Polyanilines: from solutions to polymer metal, from chemical curiosity to technology, *Synth. Met.* 69 (1995) 179.
12. Joo J, Epstein AJ, Electromagnetic radiation shielding by intrinsically conducting polymers, *Appl. Phys. Lett.* 65 (1994) 2278.
13. Joo J, Lee CY, Song HG, Kim JW, Jang KS, Oh EJ, Epstein AJ, Enhancement of electromagnetic interference shielding efficiency of polyaniline through mixture and chemical doping, *Mol. Cryst. Liq. Cryst. Sci. Technol., Sect. A: Mol. Cryst. Liq. Cryst.* 316 (1998) 367.
14. Lee CY, Song HG, Jang KS, Oh EJ, Epstein AJ, Joo J, Electromagnetic interference shielding efficiency of polyaniline mixtures and multilayer films, *Synth. Met.* 102 (1999) 1346.
15. Joo J, Lee CY, High frequency electromagnetic interference shielding response of mixtures and multilayer films based on conducting polymers, *J. Appl. Phys.* 88 (2000) 513.
16. Hong YK, Lee CY, Jeong CK, Lee DE, Kim K, Joo J, Method and apparatus to measure electromagnetic interference shielding efficiency and its shielding characteristics in broadband frequency ranges, *Rev. Sci. Instrum.* 74 (2003) 1098.
17. Kaynak A, Unsworth J, Clout R, Mohan AS, Beard GE, A study of microwave, reflection, absorption, and shielding effectiveness of conducting polypyrrole films, *J. Appl. Polym. Sci.* 54 (1994) 269.
18. Kaynak A, Mohan AS, Unsworth J, Clout R. Plane-wave shielding effectiveness studies on conducting polypyrrole, *J. Mater. Sci. Lett.* 13 (1994) 1121.
19. Kaynak A, Electromagnetic shielding effectiveness of galvanostatically synthesized conducting polypyrrole films in the 300–2000MHzfrequency range, *Mater. Res. Bull.* 31 (1996) 845.
20. Pomposo JA, Rodríguez FJ, Patent Application P9800626.
21. Husić S, Javni I, Petrović ZS, Thermal and mechanical properties of glass reinforced soy-based polyurethane composites, *Compos Sci Technol.* 65 (2005) 19–25. doi:10.1016/j. compscitech.2004.05.020.
22. Ummartyotin S, Juntaro J, Sain M, Manuspiya H, Development of transparent bacterial cellulose nanocomposite film as substrate for flexible organic light emitting diode (OLED) display, *Ind Crops Prod.* 35 (2012) 92–97. doi:10.1016/j.indcrop.2011.06.025.
23. Chen Z, Lu H, Constructing sacrificial bonds and hidden lengths for ductile graphene/polyurethane elastomers with improved strength and toughness, *J Mater Chem.* 22 (2012) 12479–12490. doi:10.1039/c2jm30517h.

24. Valentini M, Piana F, Pionteck J, Lamastra FR, Nanni F, Electromagnetic properties and performance of exfoliated graphite (EG)—thermoplastic polyurethane (TPU) nanocomposites at microwaves. *Compos Sci Technol* 114 (2015) 26–33. doi:10.1016/j.compscitech.2015.03.006.
25. Kumar A, Alegaonkar PS, Impressive transmission mode electromagnetic interference shielding parameters of graphene-like nanocarbon/polyurethane nanocomposites for short range tracking countermeasures. *ACS Appl Mater Interfaces.* 7 (2015) 14833–14842. doi:10.1021/acsami.5b03122.
26. Hsiao S-T, Ma C-CM, Tien H-W et al. Using a noncovalent modification to prepare a high electromagnetic interference shielding performance graphene nanosheet/water- borne polyurethane composite, *Carbon* 60 (2013) 57–66. doi:10.1016/j.carbon.2013.03.056.
27. Tong XC, *Advanced Materials and Design for Electromagnetic Interference Shielding*, CRC Press, Taylor and Francis Group, Boca Raton, FL, 2009.
28. Chung DDL, Electromagnetic interference shielding effectiveness of carbon materials, *Carbon.* 39 (2001) 279–285.
29. Markham D, Effective electromagnetic shield using conductive polyaniline films, *Mater. Des.* 21 (2000) 45–50.
30. Durkan C, Welland, ME, Size effects in the electrical resistivity of polycrystalline nanowires, *Phys. Rev. B: Condens. Matter Mater. Phys.* 61 (2000) 14215–14218.
31. Arjmand M, Chizari K, Krause B, Pötschke P, Sundararaj U, Impact of synthesis temperature on structure of carbon nanotubes and morphological and electrical characterization of their polymeric nanocomposites, *Carbon* 98 (2016) 358–372.
32. Chizari K, Daoud MA, Ravindran AR, Therriault D, 3D Printing of highly conductive nanocomposites for the functional optimization of liquid sensors, *Small* 12 (2016) 6076–6082.
33. Thomassin JM, Jérôme C, Pardoen T, Bailly C, Huynen I, Detrembleur C, Polymer/carbon based composites as electromagnetic interference (EMI) shielding materials, *Mater. Sci. Eng. R-Rep.* 74 (2013) 211–232.

6

ABSORPTION- AND REFLECTION-DOMINATED MATERIALS FOR ELECTROMAGNETIC INTERFERENCE SHIELDING APPLICATIONS

Lavanya Jothi

⇒ **CONTENTS**

6.1 Introduction 65
6.2 Electromagnetic Interference Shielding 66
6.3 Overview of Materials Used in Electromagnetic Shielding 66
6.4 Absorption-Based Materials in Electromagnetic Shielding 68
6.4.1 Carbon-Based Materials 68
6.4.2 Polymer-Based Materials 69
6.4.3 Other Materials 70
6.5 Reflection-Based Materials for Electromagnetic Shielding 70
6.5.1 Carbon-Based Materials 71
6.5.2 Polymer-Based Materials 72
6.5.3 Other Materials 73
6.6 Conclusion 73
6.7 Future Outlook 74
References 74

6.1 INTRODUCTION

In general, electromagnetic radiation refers to a wave that has both electric and magnetic fields perpendicular to each other, propagating through space and carrying energy-charged particles. Stars are the

DOI: 10.1201/9781003217312-6

main source of electromagnetic radiation; all other natural sources generate weak electromagnetic fields[1]. In today's materialistic world, apart from natural sources, the electromagnetic radiation that radiates from electronic appliances called electromagnetic radiation interference (EMI), is a form of electromagnetic energy that impacts the performance of electronic appliances by introducing unwanted signals or resulting in whole operational failure. As electromagnetic radiation enters the human body, heat is generated inside the body by interaction with the molecules. Hence, this leads to weakening of the nerves, insomnia, headache and languidness as the heat produced is not readily dissipated, resulting in more exposure to electromagnetic radiation[2]. Therefore, in order to increase the performance of electronic appliances and also for humans to have healthy lives, there is a need for electromagnetic interference shielding.

6.2 ELECTROMAGNETIC INTERFERENCE SHIELDING

Electromagnetic interference shielding is the method used to maintain the efficiency of electronic devices by using conductive or magnetic materials to protect against incoming or outgoing radiation at electromagnetic frequencies. The shielding effectiveness (SE) of a material denotes its capacity to perform as a protective layer against external electromagnetic fields. It is calculated as the ratio of incident energy to reflected energy. Generally, an active EMI shielding material performs three vital functions: reflection, absorption, and multiple reflections[3–5].

The reflection of EMI depends on the conductivity of the materials, when the EMI waves interact with free charge carriers like electrons or holes. Part of the waves will be reflected and another part will be absorbed. Therefore, in the case of reflection-dominant materials, the shielding efficiency is based on the incident and reflected EMI waves, whereas in the case of absorption, it is directly dependent on the thickness and the morphology of the material[6].

6.3 OVERVIEW OF MATERIALS USED IN ELECTROMAGNETIC SHIELDING

For the past few decades, metal-based materials were the most efficient materials for EMI shielding. However, limitations due to their low flexibility, high weight density, heavy processing or manipulation, and their propensity to corrosion constrained the usage of metals and metal-based composites in today's advanced electronic devices. To overcome the previously mentioned drawbacks of metal-based materials, much effort has focused on the development of nanocomposites and conductive polymer materials due to their easy processability, flexibility, low specific weight, chemical and corrosion resistance, and tunable structural and mechanical properties.

Over the past few decades, because of their high electrical conductivity (Table 6.1) and good permeability, metallic materials such as silver, copper, aluminum, and nickel have been utilized as effective shielding materials. But, in spite of their efficient EMI shielding, EMI waves have not been completely eliminated, and these materials also are disadvantageous in terms of their weight and flexibility as today's electronic devices have become faster, smaller, and lighter[7].

Meanwhile, the absorption-based EMI shielding materials, such as polymer-based and nanocomposite materials with outstanding properties like low ionization potential, high electron affinity, and high conductivity, show comparatively higher shielding efficiency than the reflection-based EMI materials. Absorption-based shielding

TABLE 6.1
Electrical Conductivity of Metals

S. No.	Metal	Electrical Conductivity (S/m)
1	Silver	6.8×10^7
2	Copper	6.4×10^7
3	Aluminum	4.0×10^7
4	Brass	1.7×10^7
5	Nickel	9.7×10^6
6	Steel	6.3×10^6
7	Stainless Steel	1.8×10^6
8	Electroless Nickel	1.8×10^6
9	Graphite	5.0×10^4
10	Gold	4.5×10^7
11	Cobalt	1.6×10^7
12	Lead	4.8×10^6
13	Iron	1.0×10^7

materials have gradually come to be preferred in many applications, such as the military or stealth technology[8].

Due to their excellent properties like flexibility, easy processability, chemical resistance, and light weight, abundant research studies have been performed on the development of polymer-based materials, which are preferable in military applications such as stealth technology and camouflage. Intrinsic conducting polymers (ICPs), such as polyaniline (PANI) and polypyrene (PPy), are common choices because of their high electrical conductivity, which can be improved through chemical doping[9].

6.4 ABSORPTION-BASED MATERIALS IN ELECTROMAGNETIC SHIELDING

Remarkable research has been performed on absorption-based EMI shielding. The efficiency of this mechanism depends on two main factors: one is the thickness of the shielding material and the other is the electric and/or magnetic dipoles present in the shield, which can interact with the EM radiation[10–13]. Materials with high permittivity and permeability like $BaTiO_3$, Fe_2O_3, ZrO_2, Fe_3O_4, ferrites, super permalloy, and mu-metal are the main sources of the electric dipoles, and magnetic dipoles are in materials with high permeability[14–16].

$$SE_A = 20 \log t/\delta \tag{6.1}$$

In the preceding equation (6.1), t denotes the thickness of the shielding material and δ denotes the skin depth. Therefore, the efficiency of absorption increases with increasing frequency of EM radiation and with increasing thickness and permeability of the shielding material. In the absence of magnetic properties, EMI shielding is exclusively dependent on the dielectric properties and vice versa.

6.4.1 Carbon-Based Materials

In 2019, Yuezhen Bin et al. reported work on buckypaper (BP) as an efficient EMI shielding material[17]. In this study, shielding efficiency (SE) above 100 dB could be achieved by constructing many layers or sandwich structures using BP and polypropylene sheet BP. In the X-band frequency range (8.2–12.4 GHz), EMI SE of 31.2 dB and a high specific SE of 19,850 dB·cm^2/g were obtained for BP of 35 μm. In the case of the multilayer structure, the enhancement in EMI SE is provided by absorption loss and results in a slow decrement on the whole with increasing BP layers. However, for the sandwich structure, the enhancement in SE is attributed to constructive interference, and there is a best polypropylene sheet thickness (0.32–0.48λ, λ = 25 mm) to realize the maximization of SE. When compared with a multilayer structure with the same components of BP, the SE of a sandwich structure with one and two wave-transmitting layers significantly increased by 61.5% and 90%, respectively.

6.4.2 Polymer-Based Materials

Due to excellent properties like the ease of synthesis, strong environmental stability, the simple regulation of conductivity by changing the oxidation and protonation states, tunable electrical and optical properties, and a special doping mechanism, the present and future semiconductor device relies on polyaniline[18]. In 2021, N. Maruthi et al. reported a study on polyaniline-coated, niobium pentoxide (PANI-Nb_2O_5) nanocomposites as potential EMI shielding materials with excellent anticorrosion behavior[19]. The fabricated nanocomposites demonstrate EMI SE in the range of –27.7 to –28.8 dB over the broadband frequency range 12–18 GHz, covering the microwave Ku-band of practical relevance. By varying the concentration of Nb_2O_5 in PANI, they observed that there was a reduction of ~99.88% with absorbed EM power of 71–80.6%.

In 2021, Jia et al. reported a study on conductive polymer composite (CPC) foams. To enhance SE, the authors fabricated designable MXene-based CPC foams with high-efficiency EMI shielding performance and ultralow reflectivity. Here, novel MXene-decorated polymer foam beads (MPFBs) were fabricated to serve as building blocks, with the assistance of polyaniline (PANI), and CPC foams were constructed by assembling MPFBs into a 3D accumulation with MXene networks, followed by the encapsulation of polydimethylsiloxane (PDMS)[17]. The fabricated material with ~0.0225–0.0449 vol% MXene (plus ~0.02 vol% PANI) showed an EMI SE of ~23.5–39.8 dB, along with a low *R* coefficient of ~0.20–0.31. The efficiency of the material was further increased by applying the asymmetric gradient configuration at total MXene content of only ~0.0225 vol% with an ultralow *R* coefficient of ~0.05, which is the lowest *R* value for effective EMI shields ever reported. Finally, in this work, the authors designed effective and green CPC foam EMI shields for the next generation of electronic devices.

6.4.3 Other Materials

In 2020, Xiaomeng Fan et al. reported a study on $Ti_3C_2T_x$[6], which are 2D inorganic compounds, as efficient EMI shielding materials, because MXenes are easier to synthesize with low cost and on a large scale. In this work, the authors synthesized $Ti_3C_2T_x$-bonded carbon black (CB) films by the VAF method and applied them as absorption-dominant EMI materials. The role of carbon black in MXenes is to introduce a porous structure into the film[20]. The authors varied the size of the pores and studied the SE of the film. Thus, they prepared $Ti_3C_2T_x$-bonded carbon black films with a porous structure. The authors observed that the reflection effectiveness (SE_R) decreased from 20 to 12 dB, and the absorption effectiveness (SE_A) increased from 31 to 47 dB. Moreover, with the increase in CB particles, the authors observed an increase in EMI SE. And the best EMI shielding effectiveness was noted to be 60 dB, with SE_A of 15 dB and SE_R of 45 dB and a specific shielding effectiveness of 8,718 dB·cm^2/g. Finally, the authors conclude that the porous structure increases the absorption of the EMI shielding films by subsequent enhanced scattering and reflection. Therefore, with this research, the author provides a potential MXene-based EMI shielding film with light weight and improved flexibility.

6.5 REFLECTION-BASED MATERIALS FOR ELECTROMAGNETIC SHIELDING

Reflection is the foremost approach of EMI shielding. This approach is based on the principle of the Faraday cage. The necessary conditions for EMI reflection are as follows.

1. The material should be conductive and is not required to reach the percolation threshold[21]. Moreover, the SE of the material increases with increasing conductivity.
2. This increase in the conductivity can be due to the material's specific property, or the conductivity of the material can be increased with an increase in the conductive filler material.
3. In most of the materials, the efficacy of reflection is enhanced with increasing shield conductivity and declines with increasing shield permeability and frequency of EM radiation.
4. The shielding efficiency of the material not only depends on conductivity but also depends on the thickness of the shielding

material. When the thickness of the shielding material is much lower than the skin depth, the reflection shielding efficiency becomes attenuated. This condition takes place at either low frequencies or, in the case of thin material, with good electrical conductivity.

5. Moreover, the efficiency of reflection also varies upon the difference in the impedance of the incident EM wave and the shielding material (equation 6.2):

$$SE_R = 20 \log[Z_o/4Z_s] \tag{6.2}$$

6.5.1 Carbon-Based Materials

In order to enhance the SE and reduce the effect of secondary EM radiation, the author has tried to stimulate the magnetic and electrical losses by using highly efficient electromagnetic shielding materials, such magnetic ferro/ferric oxide deposited on reduced graphene oxide (rGO@Fe_3O_4) and silver-coated tetra-needle-like ZnO whisker (T-ZnO/Ag) for next-generation communication technologies and high-powered electronic instruments[22]. In this research work, a new method is employed to provide flexible, waterborne, polyurethane composite films with two different nanoparticles. During film formation, a gradient structure of rGO@Fe_3O_4 and T-ZnO/Ag is automatically formed. This gradient structure plays a primary role in shielding the EMI. In this flexible film, the rGO@Fe_3O_4 takes the role of shielding the EMI via absorption, while T-ZnO/Ag offers excellent electromagnetic reflection ability for the film. The specialty of the structure is to shield the EMI in an "absorb–reflect–reabsorb" process when electromagnetic waves penetrate into the composite film, leading to excellent EMI shielding performance with an extremely low reflection characteristic even at negligible nanofiller content (0.8 vol% Fe_3O_4@rGO and 5.7 vol% T-ZnO/Ag). Finally, in this research the EMI SE reached 87.2 dB against the X-band with a thickness of only 0.5 mm, while the SE_R was only 2.4 dB and the power coefficient of reflectivity (*R*) was as low as 0.39. From the results obtained, the author proved that the flexible thin film with low nanofiller content is a potential shielding material.

The first experimental research on EMI shielding of a carbon-carbon composite was done in 1999. Here the author claims that the carbon-carbon composite shows better shielding toward EMI than continuous carbon fiber polymer-matrix composites. In this work, the carbon filament of diameter 0.1 μm is much smaller than the conventional carbon fiber and is used for improving vibration damping ability. The EMI waves incident on the composite material reflect back with shielding effectiveness of 124 dB, low surface impedance, and high reflectivity in the frequency range from 0.3 MHz to 1.5 GHz[23]. Moreover, the addition of 2.9 vol% discontinuous 0.1-μm-diameter carbon filaments between the layers of conventional 7-μm-diameter continuous carbon fibers in a composite degraded the shielding effectiveness. The dominant EMI shielding mechanism for both carbon-matrix and polymer-matrix continuous carbon fiber composites is reflection.

6.5.2 Polymer-Based Materials

Due to unique physicochemical properties like improved mechanical properties, enhanced electrical conductivity between 0.1 and 10^{-10} S/cm, and magnetic properties, polyaniline (PANI), a type of conducting polymer, is one of the polymers that is frequently used as a host material for micro- or nano-sized nanofillers. To fabricate an effective EMI shielding material, ferromagnetic materials possessing high permeability, such as Fe, Fe_3O_4, and Fe_2O_3, dielectric materials such as TiO_2, SiO_2, and ZnO, and carbonaceous materials such as graphene, multiwalled carbon nanotube, and reduced graphene oxide are widely used in PANI.

In 2018, Qiu et al. worked on the conducting polymer PANI. In this work, acid treatment with different polarities led to the self-assembly of PANI particles into fine nanostructures by preventing secondary growth[24]. The PANI treated with sulfuric acid, hydrochloric acid (HCl), or camphorsulfonic acid (CSA) displays different morphology like the nanofibers' "holothurian-like" structure. The results obtained indicated that it is possible to control the electrical and the EMI shielding properties of PANI by doping with acid under different conditions. Moreover, the absorption is dominant to the total EMI SE of 20.7 dB for PANI-CSA at a thickness of 0.35 μm.

Due to the unique properties of polycarbonate-like, high-performance, tough amorphous and transparent polymers with a unique combination of properties like high impact strength, high dimensional stability, and good electrical properties, in 2015 Gedler et al. reported research on polycarbonate-graphene composites foamed with supercritical carbon dioxide[25]. The novelty of this work relies on the 2-step foaming method, which improved graphene dispersion and led to a different cellular structure compared to that reported using 1-step foaming. In this work, in foamed composites the shielding efficiency was observed to increase 10 times compared to unfoamed composites. Moreover, the authors observed that the polycarbonate-graphene composite material achieved a maximum reflection-based EMI specific shielding effectiveness of ~78 dB·cm^3/g in foam, which was more than 70 times higher that of the unfoamed polymer (1.1 dB·cm^3/g). By controlling the foaming process conditions and nanoparticle characteristics, the authors found that it is possible to improve multiple properties while achieving lightweight materials suitable for various applications.

6.5.3 Other Materials

Due to its unique, three-dimensional (3D), porous biological structure, S. Li et al., in 2021, developed natural loofah as a shielding material with remarkable electromagnetic shielding performance[26]. In this work, the authors synthesized a carbon-coated, carbonized loofah sponge (CCLS) via the chemical vapor deposition method and directionally arrayed MXene aerogels through a simple and pollution-free freeze-drying method. CCLS with a 3D skeleton provides a mechanically strong support for the MXene aerogel and endows the composite with a good foundational network for electric conduction and shielding electromagnetic waves. The compressive strength and modulus of CCLS reached 0.979 MPa and 10.881 MPa, respectively, and an electrical conductivity of 28.34 S/m was attained after 60 h of carbon coating. The maximum conductivity and average total electromagnetic shielding effectiveness of the composites prepared achieved levels of 55.99 S/m and 70.0 dB, respectively. Moreover, the electromagnetic shielding effectiveness value of the composites could still reach 42.6 dB after being burned in an alcohol flame for 120 s.

6.6 CONCLUSION

For the past few decades, the extensive increase in the need for and use of electronic equipment has led to very high exposure to EMI in day-to-day life. Because the available materials existing in the field were made up of metals and had disadvantages such as heavy weight and high cost, a significant amount of research was directed toward finding novel materials to shield EMI over a broad frequency range, that is, in the range of MHz to GHz. In brief, this chapter provided detailed insight into absorption- and reflection-based electromagnetic shielding materials. After a short introduction on the need for shielding materials, the materials that absorb and reflect were closely described.

6.7 FUTURE OUTLOOK

The following future research ideas should be explored to shield EMI:

1. Because the porous structured materials show excellent efficiency, a porous structured carbon nanotube-MXene composite is expected to be an excellent selection for shielding EMI not only in the MHz range but also in the GHz range.
2. Polymerization of the surface functional group over the filler can be done by a plasma functionalization technique. This specific functionalization method will be helpful for maintaining the morphology and controlling the functionalization of materials for better distribution of the filler than bulk mixing in the composite.

REFERENCES

1. Jauchem, J. R. Exposure to extremely-low frequency electromagnetic fields and radiofrequency radiation: Cardiovascular effects in humans. *Int. Arch. Occup. Environ. Health.* **70**, 9–21 (1997).
2. Keangin, P., Vafai, K. & Rattanadecho, P. Electromagnetic field effects on biological materials. *Int. J. Heat Mass Transf.* **65**, 389–399 (2013).
3. Elmas, O. Effects of electromagnetic field exposure on the heart: A systematic review. *Toxicol. Ind. Health.* **32**, 76–82 (2013).
4. Kim, H.-R., Fujimori, K., Kim, B.-S. & Kim, I.-S. Lightweight nanofibrous EMI shielding nanowebs prepared by electrospinning and metallization. *Compos. Sci. Technol.* **72**, 1233–1239 (2012).
5. Al-Saleh, M. H., Saadeh, W. H. & Sundararaj, U. EMI shielding effectiveness of carbon based nanostructured polymeric materials: A comparative study. *Carbon N. Y.* **60**, 146–156 (2013).
6. Shahzad, F. *et al.* Electromagnetic interference shielding with 2D transition metal carbides (MXenes). *Science.* **353**, 1137–1140 (2016).
7. Wanasinghe, D. & Aslani, F. A review on recent advancement of electromagnetic interference shielding novel metallic materials and processes. *Compos. Part B Eng.* **176**, 107207 (2019).

8. Geetha, S., Satheesh Kumar, K. K., Rao, C. R. K., Vijayan, M. & Trivedi, D. C. EMI shielding: Methods and materials—A review. *J. Appl. Polym. Sci.* **112**, 2073–2086 (2009).
9. Thomassin, J.-M. *et al.* Polymer/carbon based composites as electromagnetic interference (EMI) shielding materials. *Mater. Sci. Eng. R Reports.* **74**, 211–232 (2013).
10. Chen, L. *et al.* Mechanical and electromagnetic shielding properties of carbon fiber reinforced silicon carbide matrix composites. *Carbon N. Y.* **95**, 10–19 (2015).
11. Cao, M.-S. *et al.* 2D MXenes: Electromagnetic property for microwave absorption and electromagnetic interference shielding. *Chem. Eng. J.* **359**, 1265–1302 (2019).
12. Duan, J., Wang, X., Li, Y. & Liu, Z. Effect of double-layer composite absorbing coating on shielding effectiveness of electromagnetic shielding fabric. *Mater. Res. Express.* **6**, 86109 (2019).
13. Jia, Z. *et al.* Graphene foams for electromagnetic interference shielding: A review. *ACS Appl. Nano Mater.* **3**, 6140–6155 (2020).
14. Ma, Z. *et al.* The influence of different metal ions on the absorption properties of nano-nickel zinc ferrite. *Materials.* **11** (2018).
15. Dong, C. *et al.* Microwave magnetic and absorption properties of M-type ferrite BaCoxTixFe12−2xO19 in the Ka band. *J. Magn. Magn. Mater.* **354**, 340–344 (2014).
16. Li, Y. *et al.* Multifunctional BiFeO3 composites: Absorption attenuation dominated effective electromagnetic interference shielding and electromagnetic absorption induced by multiple dielectric and magnetic relaxations. *Compos. Sci. Technol.* **159**, 240–250 (2018).
17. Jia, X., Shen, B., Zhang, L. & Zheng, W. Construction of compressible polymer/MXene composite foams for high-performance absorption-dominated electromagnetic shielding with ultra-low reflectivity. *Carbon N. Y.* **173**, 932–940 (2021).
18. Li, X.-G., Li, A. & Huang, M.-R. Facile high-yield synthesis of polyaniline nanosticks with intrinsic stability and electrical conductivity. *Chemistry.* **14**, 10309–10317 (2008).
19. Maruthi, N. *et al.* Promising EMI shielding effectiveness and anticorrosive properties of PANI-Nb2O5 nanocomposites: Multifunctional approach. *Synth. Met.* **275**, 116744 (2021).
20. Fan, X. *et al.* Electromagnetic interference shielding Ti3C2Tx-bonded carbon black films with enhanced absorption performance. *Chinese Chem. Lett.* **31**, 1026–1029 (2020).
21. Zhang, D.-Q. *et al.* Self-assembly construction of WS2—rGO architecture with green EMI shielding. *ACS Appl. Mater. Interfaces.* **11**, 26807–26816 (2019).
22. Xu, Y. *et al.* Gradient structure design of flexible waterborne polyurethane conductive films for ultraefficient electromagnetic shielding with low reflection characteristic. *ACS Appl. Mater. Interfaces.* **10**, 19143–19152 (2018).

23. Luo, X. & Chung, D. D. L. Electromagnetic interference shielding using continuous carbon-fiber carbon-matrix and polymer-matrix composites. *Compos. Part B Eng.* **30**, 227–231 (1999).
24. Qiu, M., Zhang, Y. & Wen, B. Facile synthesis of polyaniline nanostructures with effective electromagnetic interference shielding performance. *J. Mater. Sci. Mater. Electron.* **29**, 10437–10444 (2018).
25. Gedler, G., Antunes, M., Velasco, J. I. & Ozisik, R. Electromagnetic shielding effectiveness of polycarbonate/graphene nanocomposite foams processed in 2-steps with supercritical carbon dioxide. *Mater. Lett.* **160**, 41–44 (2015).
26. Li, S. *et al.* CVD carbon-coated carbonized loofah sponge loaded with a directionally arrayed MXene aerogel for electromagnetic interference shielding. *J. Mater. Chem. A.* **9**, 358–370 (2021).

7

HIGH-TEMPERATURE MATERIALS FOR ELECTROMAGNETIC INTERFERENCE SHIELDING APPLICATIONS

Ragin Ramdas M.

⇒ **CONTENTS**

7.1 Introduction 77
7.2 Silicon-Based Materials 78
7.3 Carbon-Based Materials 80
7.4 Shape Memory Materials 81
7.5 Other Materials 82
7.6 Conclusion 82
References 83

7.1 INTRODUCTION

Electromagnetic interference or EMI is the interference caused by one electronic device with another through the electromagnetic field set up by its operation. This can arise from many sources in diverse modes. The various types of EMI can be categorized thus:

1. Man-made EMI: This type generally arises from other electronic circuits, although some could be from the switching of large currents.
2. Naturally occurring EMI: Cosmic noise, lightning, and other atmospheric types of noises can all contribute as sources of this type of EMI.
3. Continuous interference: This stems from sources such as circuits that emit a continuous signal. However, the background noise, which is continuous, may be created in natural or man-made ways.

DOI: 10.1201/9781003217312-7

4. Impulse noise: Lightning, electrostatic discharge (ESD), and switching systems all contribute to impulse noise, which is a type of man-made or natural EMI.

EMI is present in all areas of electronics. An ideal comprehension of the source, coupling methods, and victim susceptibility enable the reduction of interference to such a level wherein the EMI causes no undue degradation in performance. Today, EMI shielding materials have garnered immense attention around the world. This is primarily due to their extensive application in protecting electronic devices from electromagnetic interference[1–3]. The development of the first EMI absorbing shield based on carbon black (CB) and titanium dioxide was patented in The Netherlands in 1936[4]. Two popular types of EMI shielding materials were metal-based composites and high conducting polymer–based composites. However, metal corrodes easily and the reflected waves interfere with electronic devices in its vicinity, making them poor materials for EMI shielding applications. The other main drawbacks of metals are their limited physical flexibility, higher density, and material cost. Ceramics, on the other hand, due to their low density and high corrosion resistance, can be applied as high-temperature structural materials, but their low electrical conductivity renders them ineffective as EMI shielding materials.

Different from conventional ceramics, MAX phases have high electrical conductivity, similar to that of metals. MXenes are layered materials made from the precursor MAXene by removing the nonmetallic (A) element. This is due to the M-X metallic bonding within the lattice structure, which reveals their superior application potential as EMI shielding materials[5–14].

Metal organic frameworks (MOFs) are well-ordered crystals with coordinated metal ions and organic molecules, and MOF-derived magnetic metal nanoparticle/carbon composites possess dielectric and magnetic loss, a large surface area, and are lightweight, which provide excellent EM wave attenuation performance[15–17].

7.2 SILICON-BASED MATERIALS

SiC-derived composites have been used as good microwave absorption materials due to their high strength, low oxidation, high thermal stability, and thermal conductivity at higher temperatures. But the main drawback of SiC-based composites is their lower dielectric and absorption properties[18]. SiC-C solid solution material systems were obtained by annealing the polycarbosilane-derived SiC at ultrahigh temperatures. These serve as EMI shielding materials because of the unique distribution states of free carbon formed *in situ* in the materials. At annealing temperatures of 1,700, 1,900, and 2,000 °C, the average total shielding effectiveness per unit thickness was 29.14, 18.05, and 22.08 dB/mm, respectively[19]. Ordered, mesoporous, interfilled SiC/SiO_2 monolithic composites were prepared via nanocasting and cold-pressing, using polycarbosilane as the precursor and mesoporous silica SBA-15 as the structural template. Their high-temperature dielectric properties and microwave attenuation performance were evaluated in the temperature range of 25–500 °C in the X band (8.2–12.4 GHz). These composites exhibited high-temperature microwave absorbing properties, and attenuation mechanisms were proposed[20].

Ti_3SiC_2 filler was introduced into SiC_f/SiC composites through precursor infiltration and pyrolysis (PIP). This was to optimize their dielectric properties for EMI shielding applications at temperatures ranging from 25 to 600 °C at 8.2–12.4 GHz. After the filler was incorporated, the flexural strength of the SiC_f/SiC composites improved from 217 to 295 MPa. Their temperature-dependent behavior is highlighted through the complex permittivity and tan δ of the composites, which increase with increasing temperature. The absorption, reflection, and total shielding effectiveness of the composites with Ti_3SiC_2 filler are enhanced from 13, 7, and 20 dB to 24, 21, and 45 dB, respectively, when the temperature increases from 25 to 600 °C[21]. The EMI shielding of dense Ti_3AlC_2 ceramics with distinct microstructures was investigated up to 800 °C. Ti_3AlC_2 exhibited a high EMI shielding effectiveness (SE) of about 30 dB, which was mainly attributed to its high electrical conductivity and complex permittivity. In addition, the EMI shielding was adequately sensitive to the microstructure. Ti_3AlC_2 with high aspect ratio grains and a certain degree of texture was preferred for high SE, indicating that typically layered structures also contribute to the high shielding capability. Thus, layered Ti_3AlC_2 ceramics could be promising structural EMI shielding materials at high temperatures[22].

Lightweight and flexible ZrC/SiC hybrid nanofiber mats were successfully fabricated by electrospinning and high-temperature pyrolysis with polycarbosilane (PCS) and zirconium acetylacetone [$Zr(acac)_3$] as precursors. The addition of $Zr(acac)_3$ renders a PCS solution of lower viscosity and higher conductivity. After pyrolysis, the highly conductive ZrC nanoparticles are uniformly distributed within the SiC nanofiber matrix. Hence, the average diameter of the

ZrC/SiC nanofibers is reduced from 2.6 μm to 330 nm compared to pure SiC fibers. It was noted that 3-layered ZrC/SiC nanofiber mats with a total thickness of 1.8 mm achieved an EMI shielding effectiveness (SE_T) of 18.9 dB. At a low density of 0.06 g/cm^3, the specific shielding effectiveness could be as high as 315 dB·cm^3/g. Moreover, the SE_T value could be further enhanced to 20.1 dB at 600 °C[23].

7.3 CARBON-BASED MATERIALS

Carbon-based polymer composites have gained popularity recently because of the combination of their light weight, resistance to corrosion, flexibility, electrical conduction, and processing advantages. Different types of nanowires, including multiwalled carbon nanotubes coated with CdS nanocrystals or coated with different thickness CdS sheaths, have been synthesized through mild solution process synthesis. The composite loading with 6 vol% CdS-MWCNTs shows the best absorption of −47 dB at 473 K, with a thickness of 2.6 mm in the temperature range of 323–573 K and in the X-band frequency region[24]. Electrically conducting, thermally reduced graphene (TRG) nanosheets were synthesized through thermal exfoliation and subsequent annealing of grapheme oxide at 800 °C. Thermoplastic polyurethane–based (TPU) nanocomposites with different concentrations (ranging between 0 and 5.5 vol%) of TRG nanosheets were prepared by the solution blending method. A TPU/TRG nanocomposite at 5.5 vol% loading exhibits an enhanced electrical conductivity on the order of 3.1×10^{-2} S/m and shows a superior EMI SE of approximately –26 to –32 dB in the Ku-band frequency region[25].

A high-performance EMI shielding material based on carbon nanofibers (CNFs) and cellulose filter (CF) paper was fabricated through a cost-efficient, convenient dip-coating method. The EMI shielding performance of the CF papers with micron-level thickness tolerance (2.5–12.7 μm) has been explored by considering the microstructure, serviceability, electrical conductivity, and number of dip-coating cycles. Field emission scanning electron microscopy of the surface and edge of the composites supports good electrical conductivity, with a distinct increase from 6.6×10^{-7} to 0.85 S/cm. After the dipping cycle was altered, it was concluded that the number of dip-coating cycles significantly impacts electrical conductivity. The electromagnetic shielding efficiency of CNF-coated CF paper was 24.6 dB with only 25 dip-coating cycles. Moreover, from a commercially viable perspective, extensive studies have investigated CNF-coated

CF papers in simulated aging environments, namely, water, thermal aging, and thermodegradability over a wide range of temperatures (ambient to 600 °C). As conductive CNF-coated CF papers possess significantly higher mechanical properties than pure CF paper, they are ideal as a highly flexible, lightweight, and cost-efficient EMI shielding material in advanced multifunctional application areas[26].

The carbon/red mud hybrid foams with superior EMI shielding and excellent fire-resistant properties were fabricated via the PU foam template route. When explored within the X-band (8.2–12.4 GHz) region, these revealed fire resistance in alcohol flame and thermal stability up to 1000 °C in an oxidative environment. Incorporation of the red mud into the carbon matrix not only improves its EMI shielding properties but also enhances its flame resistance and thermal stability. The CF-RM20 possesses high dielectric and magnetic properties with improved EMI shielding of 51.4 dB at 8.2 GHz, which is 127% higher than CF. Further, the incorporation of red mud content into carbon foam provides interfacial polarization, eddy current loss, magnetic loss, and open-cell interconnected networks, which cause high surface area and thus improve the absorption by 250% in CF-RM20. The present investigation offers a novel way to fabricate carbon/red mud hybrid foams by utilizing red mud (industrial waste), which can be used as effective shielding and fire-resistant materials for defense and aerospace applications[27]. Metal oxides like Fe_3O_4, ZnO, and MnO_2 substituted as nanoparticles on CNTs along with SiO_2 can also serve as good microwave absorption materials at high temperatures[28].

7.4 SHAPE MEMORY MATERIALS

By incorporating shape memory effects, an EMI shielding material can develop a simple way to modulate its EMI shielding performance through easy-to-access stimuli, such as heating. Electromagnetic shielding, shape memory polyimide (EMSMPI) is synthesized by incorporating 5% short carbon fiber and 4% carbon black into a shape memory polyimide matrix. Its glass transition temperature of 308 °C is higher than that of other electromagnetic shielding polymers, and it can hang a kettle 43,000 times heavier than itself. The 0.35-mm-thick EMSMPI film exhibits good shielding effects with an average EMI shielding effectiveness of 23.9 dB in the X-band region. The shape processability of EMSMPI, endowed by shape memory effects, makes it valuable for sophisticated devices as it can be easily processed into different shapes. Its shielding effectiveness even 30 thirty cycles is still higher than the recommended limit of 20 dB for commercial applications[29].

7.5 OTHER MATERIALS

β-MnO_2 nanorods of length 2–4 μm and diameter 40–100 nm with highly sufficient shielding effectiveness greater than 20 dB were synthesized via a hydrothermal method from the precursors manganese sulfate monohydrate and ammonium persulfate[30]. Plasma-sprayed ZrB_2/Al_2O_3 ceramics displayed high dielectric losses in terms of a high imaginary part of the complex permittivity ranging from 120 to 150 F/m, indicative of exceptional EMI shielding performance. The total EMI shielding efficiency (SE) of 30 wt% ZrB_2-filled Al_2O_3 ceramics is more than 31 dB at 25 °C and 44 dB at 600 °C, in the range from 8.2 to 12.4 GHz with 1.2-mm thickness. In addition, the ZrB_2/Al_2O_3 ceramics still maintain a high EMI SE and long life at high temperatures (600 °C for 300 h). Therefore, ZrB_2/Al_2O_3 ceramics may be ideal as high-temperature EMI shielding materials with their excellent shielding efficiency and long life[31].

7.6 CONCLUSION

In this chapter, materials based on silicon and carbon, shape memory–induced materials, and some other materials for EMI shielding at high temperatures were detailed. Silicon-based materials, including MAX phase materials, exhibit excellent total shielding effectiveness due to their layered structure. Of those materials, SiC_f/SiC composites with Ti_3SiC_2 and Ti_3AlC_2 fillers, namely, ZrC/SiC and SiC/SiO_2, are promising candidates with very efficient operating temperatures as high as 600–1400 °C. Carbon nanomaterials, hybrids, and their composites also can contribute a good portion toward the elimination of electromagnetic interference. Among them, CdS-MWCNTs displayed an absorption of −47 dB at 473 K, in the temperature range 323–573 K and in the X-band frequency region. A TPU/TRG nanocomposite with enhanced electrical conductivity is superior and shows an EMI SE of approximately –26 to –32 dB. Carbon nanofiber (CNF) and cellulose filter (CF) paper and carbon/red mud hybrid foams exhibit thermal stability above 600 °C. In the class of shape memory materials, EMSMPI with glass transition temperature 308 °C shows excellent shielding effectiveness even after 30 cycles. β-MnO_2 nanorods and plasma-sprayed ZrB_2/Al_2O_3 ceramics also feature shielding effectiveness greater than 20 dB. In any case, numerous combinations of materials should, however, be investigated for use in different domains including defense and aviation. An exclusive and wide research should be conducted to develop new functional materials that can overcome the existing drawbacks for future prospects for EMI shielding purposes at high temperatures.

REFERENCES

1. Chen, Y.; Zhang, H.B.; Yang, Y.; Wang, M.; Cao, A.; Yu, Z.Z. High-performance epoxy nanocomposites reinforced with three-dimensional carbon nanotube sponge for electromagnetic interference shielding. *Adv. Funct. Mater.* 2016, 26, 447–455.
2. Yin, X.; Xue, Y.; Zhang, L.; Cheng, L. Dielectric, electromagnetic absorption and interference shielding properties of pour yttria-stabilized zirconia/silicon carbide composites. *Ceram. Int.* 2012, 38, 2421–2427.
3. Xu, F.; Chen, R.; Lin, Z.; Qin, Y.; Yuan, Y.; Li, Y.; Zhao, X.; Yang, M.; Sun, X.; Wang, S.; et al. Superflexible interconnected graphene network nanocomposites for high-performance electromagnetic interference shielding. *J. Am. Chem. Soc.* 2018, 3, 3599–3607.
4. Machinerieen, N.V. Device and procedure for Improvement of production and reception devices of ultrashort electric waves. *French Patent.* 1936, 802, 728.
5. Shi, S.; Zhang, L.; Li, J. Ti3SiC2 material: An application for electromagnetic interference shielding. *Appl. Phys. Lett.* 2008, 93, 172903.
6. Shi, S.L.; Zhang, L.Z.; Li, J.S. Complex permittivity and electromagnetic interference shielding properties of Ti3SiC2/polyaniline composites. *J. Mater. Sci.* 2009, 44, 945–948.
7. Tan, Y.; Luo, H.; Zhang, H.; Zhu, X.; Peng, S. High-temperature electromagnetic interference shielding of layered Ti3AlC2 ceramics. *Scr. Mater.* 2017, 134, 47–51.
8. Li, S.; Tan, Y.; Xue, J.; Liu, T.; Zhou, X.; Zhang, H. Electromagnetic interference shielding performance of nano-layered Ti3SiC2 ceramics at high-temperatures. *AIP Adv.* 2018, 8, 015027.
9. Tan, Y.; Luo, H.; Zhou, X.; Peng, S.; Zhang, H. Dependences of microstructure on electromagnetic interference shielding properties of nano-layered Ti3AlC2 ceramics. *Sci. Rep.* 2018, 8, 7935.
10. Ma, Y.; Yin, X.; Fan, X.; Wang, L.; Greil, P.; Travitzky, N. Near-net-shape fabrication of Ti3SiC2-based ceramics by three-dimensional printing. *Int. J. Appl. Ceram. Technol.* 2015, 12, 71–80.

11. Ma, X.; Yin, X.; Fan, X.; Sun, X.; Yang, L.; Ye, F.; Cheng, L. Microstructure and properties of dense Tyranno-ZMI SiC/SiC containing Ti3Si(Al)C2 with plastic deformation toughening mechanism. *J. Eur. Ceram. Soc.* 2018, 38, 1069–1078.
12. Fan, X.; Yin, X.; Chen, L.; Zhang, L.; Cheng, L.; Zhou, Y. Mechanical behavior and electromagnetic interference shielding properties of C/SiC-Ti3Si(Al)C2. *J. Am. Ceram. Soc.* 2016, 99, 1717–1724.
13. Dong, N.; Chen, L.; Yin, X.; Ma, X.; Sun, X.; Cheng, L.; Zhang, L. Fabrication and electromagnetic interference shielding effectiveness of Ti3Si(Al)C2 modified Al2O3/SiC composites. *Ceram. Int.* 2016, 42, 9448–9454.
14. Fan, X.; Yin, X.; Cai, Y.; Zhang, L.; Cheng, L. Mechanical and electromagnetic interference shielding behaviour of C/SiC composites containing Ti3SiC2. *Adv. Eng. Mater.* 2017, 20, 1700590.
15. Zhang, Y.; Zhang, H.B.; Wu, X.; Deng, Z.; Zhou, E.; Yu, Z.Z. Nanolayered cobalt@ carbon hybrids derived from metal—organic frameworks for microwave absorption. *ACS Appl. Nano Mater.* 2019, 2(4), 2325–2335.
16. Wang, K.; Chen, Y.; Tian, R.; Li, H.; Zhou, Y.; Duan, H.; Liu, H. Porous Co—C core—shell nanocomposites derived from Co-MOF-74 with enhanced electromagnetic wave absorption performance. *ACS Applied Mater. Interfaces.* 2018, 10(13), 11333–11342.
17. Liao, Q.; He, M.; Zhou, Y.; Nie, S.; Wang, Y.; Hu, S., . . . Tong, Y. Highly cuboid-shaped heterobimetallic metal—organic frameworks derived from porous Co/ZnO/C microrods with improved electromagnetic wave absorption capabilities. *ACS Appl. Mater. Interfaces.* 2018, 10(34), 29136–29144.
18. Gogoi, J.P.; Shabir, A. High-temperature electromagnetic interference shielding materials. In *Materials for Potential EMI Shielding Applications* (pp. 379–390). Netherlands: Elsevier, 2020.
19. Jia, Y.; Chowdhury, M.A.R.; Xu, C. Electromagnetic property of polymer derived SiC—C solid solution formed at ultra-high temperature. *Carbon.* 2020, 162, 74–85.
20. Yuan, X.; Cheng, L.; Guo, S.; Zhang, L. High-temperature microwave absorbing properties of ordered mesoporous inter-filled SiC/SiO2 composites. *Ceramics Interfaces.* 2017, 43(1), 282–288.
21. Mu, Y.; Zhou, W.; Wan, F.; Ding, D.; Hu, Y.; Luo, F. High-temperature dielectric and electromagnetic interference shielding properties of SiCf/SiC composites using Ti3SiC2 as inert filler. *Compos. Part A: Appl. Sci. Manuf.* 2015, 77, 195–203.
22. Tan, Y.; Luo, H.; Zhang, H.; Zhou, X.; Peng, S. High-temperature electromagnetic interference shielding of layered Ti3AlC2 ceramics. *Scr. Mater.* 2017, 134, 47–51.

23. Hou, Y.; Cheng, L.; Zhang, Y.; Du, X.; Zhao, Y.; Yang, Z. High temperature electromagnetic interference shielding of lightweight and flexible ZrC/SiC nanofiber mats. *Chem. Eng. J.* 2021, 404, 126521.
24. Lu, M.; Wang, X.; Cao, W.; Yuan, J.; Cao, M. Carbon nanotube-CdS core—shell nanowires with tunable and high-efficiency microwave absorption at elevated temperature. *Nanotechnology.* 2015, 27(6), 065702.
25. Bansala, T.; Joshi, M.; Mukhopadhyay, S.; Doong, R.A.; Chaudhary, M. Electrically conducting graphene-based polyurethane nano-composites for microwave shielding applications in the Ku band. *J. Mater. Sci.* 2017, 52(3), 1546–1560.
26. Mondal, S.; Ganguly, S.; Das, P.; Bhawal, P.; Das, T.K.; Nayak, L.; Das, N.C. High-performance carbon nanofiber coated cellulose filter paper for electromagnetic interference shielding. *Cellulose.* 2017, 24(11), 5117–5131.
27. Kumar, R.; Sharma, A.; Pandey, A.; Chaudhary, A.; Dwivedi, N.; Mondal, D.P.; Srivastava, A.K. Lightweight carbon-red mud hybrid foam toward fire-resistant and efficient shield against electromagnetic interference. *Sci. Rep.* 2020, 10(1), 1–12.
28. Jia, Z.; Lin, K.; Wu, G.; Xing, H.; Wu, H. Recent progresses of high-temperature microwave-absorbing materials. *Nano.* 2018, 13(06), 1830005.
29. Kong, D.; Li, J.; Guo, A.; Xiao, X. High temperature electromagnetic shielding shape memory polymer composite. *Chem. Eng. J.* 2020, 127365.
30. Song, W.L.; Cao, M.S.; Hou, Z.L.; Lu, M.M.; Wang, C.Y.; Yuan, J.; Fan, L.Z. Beta-manganese dioxide nanorods for sufficient high-temperature electromagnetic interference shielding in X-band. *Appl. Phy. A.* 2014, 116(4), 1779–1783.
31. Qing, Y.; Yao, H.; Li, Y.; Luo, F. Plasma-sprayed ZrB2/Al2O3 ceramics with excellent high temperature electromagnetic interference shielding properties. *J. Eur. Ceram. Soc.* 2021, 41(2), 1071–1075.

8

ELECTROMAGNETIC INTERFERENCE SHIELDING MATERIALS FOR AEROSPACE APPLICATIONS

Ananthu Prasad, Miran Mozetič, and Sabu Thomas

⇒ **CONTENTS**

8.1 Introduction 87
8.2 Radiation in Space 88
8.3 Electromagnetic Interference in Aerospace 90
8.4 Shielding Materials for Aerospace Applications 91
8.4.1 Metal-Based EMI Shielding 91
8.4.2 Polymer Composites 93
8.4.2.1 Metal-Coated Polymers 93
8.4.2.2 Conducting Polymer–Based Materials 94
8.4.2.3 Nanofiller-Based Composites 94
8.5 Conclusion 95
References 95

8.1 INTRODUCTION

The rapid development of technology in the last century has resulted in the incorporation of electronic devices in all areas of technology. Airplanes and spacecraft became safer and more efficient with these devices. But these developments resulted in a critical problem known as electromagnetic interference (EMI). EMI is an undesirable electromagnetic wave that causes problems with electronic apparatus such as sensors, power and power subsystem units, batteries, rocket payloads, communication units, remote sensing instruments, data handling units, televisions, mobile phones, computers, underground transformers, various medical devices, military planes, space systems, and the externally located units of spacecraft and commercial airplanes [1]. In the case of

DOI: 10.1201/9781003217312-8

airplanes and spacecraft, EMI can originate from inside the electronic devices inside them or from transistors on the ground [2–10].

EMI can also arise from self-regulating phenomena, such as thunder, solar flares, and electrostatic discharge (ESD). Faraday's cage mechanism can be utilized effectively to deal with EMI. But it is difficult to develop materials that have desirable properties, such as high EMI shielding performance, weather resistance, mechanical strength, high density, simple processing, and cheap cost. The advancements in technology demand the study and development of EMI shielding materials due to the ever-increasing use of electronic devices such as personal computers, mobile phones, integrated circuits, satellites, and airplane components. Thus, the study and development of EMI shielding materials have been a prime area of concern at both the academic and industrial levels. EMI shielding materials find application in equipment and devices in various fields such as aerospace and aeronautics, electrical, electronics, military, communication, and home appliances. Presently, efficient EMI shielding materials are available in the design of metal foams, metal-polymer composites, and conducting polymer composites. The introduction of nanoparticles resulted in multifunctional shielding materials that found use in several applications, such as airframes, communication, entertainment surveillance, antistatic dissipation, ESD protection, and EMI shielding materials [10–22]. The use of shielding materials in aerospace applications demands good shielding performance even at extreme conditions. Multidisciplinary characterization enables us to estimate the performance of shielding materials in such conditions. Modern space components are designed to have good chemical and physical properties, along with the ability to provide good shielding properties without extra cost. This chapter discusses the radiation environment in aerospace applications and materials used to overcome EMI in detail.

8.2 RADIATION IN SPACE

The main challenge in designing EMI shielding materials for aerospace applications is to understand the radiation environment. Better shielding capability is required to counter high-energy electromagnetic waves. The developments in shielding materials are focused on creating mechanically stable and reliable shielding in harsh conditions. A spacecraft encounters radiation from different sources in space. A schematic representation of different sources of EMI in space is shown in Figure 8.1.

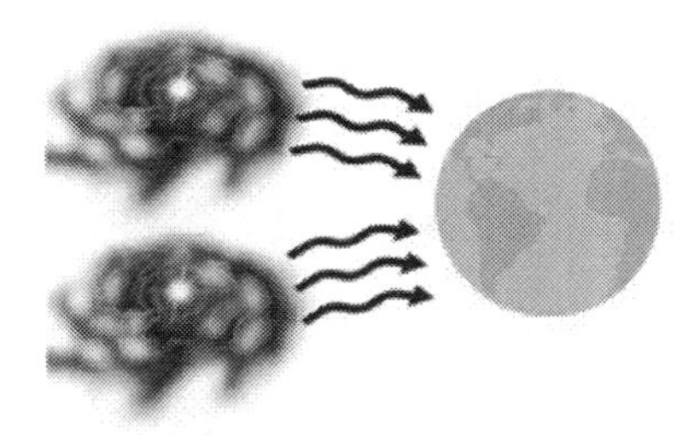

(a) Radiation from high energetic charged Supernova clouds

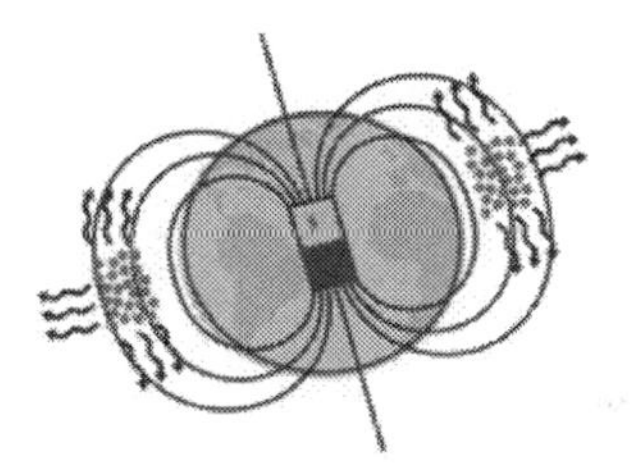

(b) Radiation due to the trapping of energetic charged particles in earth's magnetic field

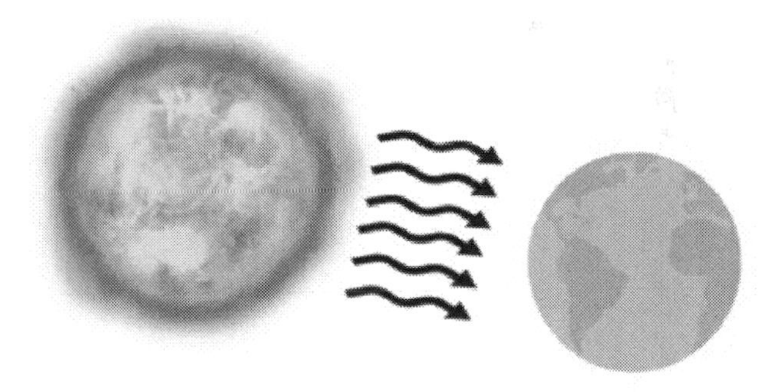

(c) Radiation from Sun

Figure 8.1 Schematic illustration of various types of radiation in space: (a) galactic cosmic radiation, (b) trapped radiation, and (c) solar energetic particles.

The universe consists of 70% vacuum, 26% dark matter, 4% common matter, and 0.005% radiation. Radiation is defined as a traveling form of energy. Electromagnetic radiation contains a stream of atomic and subatomic particles, such as protons, neutrons, electrons, alpha particles, and heavy primary nuclei, which moves with high energy and velocity. Electrons, protons, gamma rays, alpha particles, and beta particles are the primary forms of radiation. Secondary radiation is emitted when primary radiation is incident on matter. When radiation is incident upon spacecraft, in addition to secondary radiation, it causes electromagnetic interference (EMI) [23,24].

8.3 ELECTROMAGNETIC INTERFERENCE IN AEROSPACE

EMI shielding in aerospace applications has been a field of extensive research and development throughout the past decade. Due to the rapid developments in electronics and technology, scientists have faced problems with devices caused by EMI. As a result, numerous restrictions were put into place in aerospace, electronics, and communication components such as transmitters, circuit boards, and computer accessories. This forced the research and development of their EMI shielding characteristics. EMI shielding is especially important

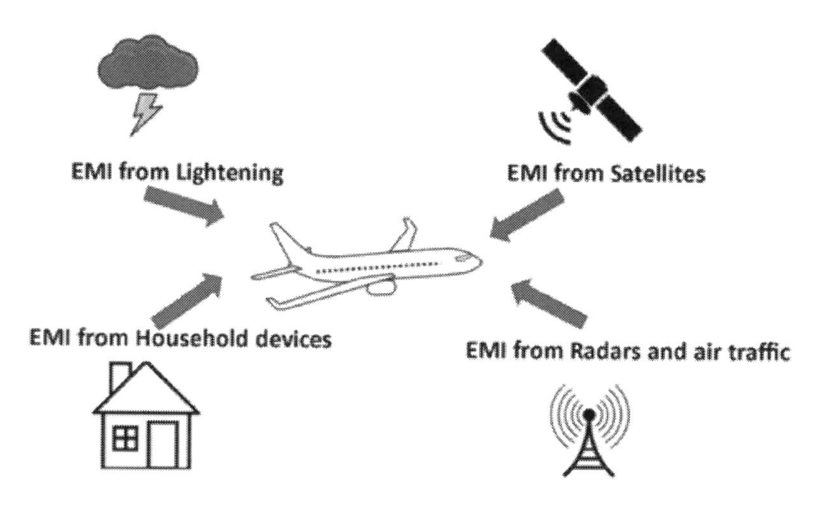

Figure 8.2 Schematic diagram of different sources of EMI in aerospace.

in airplanes and spacecraft, because energetic radiation can infiltrate through them and disturb their electric circuits, sensors, and power units [2–5]. The various sources of EMI in aerospace are shown schematically in Figure 8.2.

8.4 SHIELDING MATERIALS FOR AEROSPACE APPLICATIONS

A wide variety of shielding materials have been developed in recent years. The use of a specific material depends on the shielding environment and the application. The primary class of EMI shielding materials now in use is metal-based EMI shielding materials and polymer composites. We will discuss these two types of shielding materials in detail.

8.4.1 Metal-Based EMI Shielding

Basically, metal conductors are used to ground electronic devices, which protects them from dissipating heat and static charges [25]. The conductivity of metal is also made use of in EMI shielding. EMI shielding can be classified as active or passive. Active shielding is used for high-energy shielding requirements where pulses of charge are given to the metal, which cancels the EMI from the incident radiation. Passive shielding depends on the material properties alone

TABLE 8.1
Properties and Application of Metals for EMI Shielding [1]

Metal	Properties	EMI Shielding
Iron	Tremendous plasticity, toughness, and weld ability, great pressure processing, low strength, high iron oxide corrosion	Iron wire mesh and enclosure at 10 GHz up to 60 dB
Steel	High tensile strength and low cost, low corrosion, galvanized steels are used to prevent corrosion, higher density	Low-frequency RF shielding, low-carbon steel shows better DC and low-frequency shielding (due to high permeability and saturation point)

(Continued)

TABLE 8.1
(Continued)

Metal	Properties	EMI Shielding
Aluminum	High corrosion resistance, lightweight, high electrical conductivity, low impact resistance	Excellent shielding effectiveness of up to 100 dB
Copper, brass	Good electrical and thermal conductivity, suffers from galvanic corrosion near the contact area, nonferrous characteristics, high cost, low tensile strength, poor resistance to abrasion and common acids, better than steel	Radio frequency interference shielding, marine applications, not good for lightning or cosmic radiation shielding at 10 GHz, >60 dB
Copper-iron alloy	Excellent electromagnetic properties, electrically conducting, cost, and processing	Tremendous electromagnetic shielding performance
Magnesium	Reflection attenuation of the incident electromagnetic wave, thermal conductivity properties, shock absorber	Superior electromagnetic interference (EMI) shielding
Nickel filaments	Superior magnetic permeability and oxidation-resistant properties, better than copper	Excellent shielding effectiveness
Lead	Malleable, soft, and corrosion resistant, high density, lighter, health and environmental concerns	X-ray and gamma-ray shielding, radiation protection
Tin	Moderate oxidation compared with copper, galvanized steel, steel, aluminum, low oxidation, and high conductivity	Electromagnetic shielding material in the corrosive environment, better shielding performance

and is used in comparatively low-frequency electrical fields. Many metal-based enclosures have been developed for EMI shielding, which include single-layer metal shields, multilayer shields, double-isolated metal sheets separated by an insulating material, and shields consisting of circular or square apertures separated by a space (also known as perforated sheets) [26]. However, materials with high specific gravity are not preferred in several applications. Lightweight, flexible, and efficient absorbing materials with good absorption over a wide range are in high demand. The properties and application of different metals are shown in Figure 8.1.

8.4.2 Polymer Composites

Polymer composites are very versatile materials that have proven to be efficient in EMI shielding applications. Composite refers to a material consisting of two or more constituents. One acts as the matrix and other acts as reinforcement. Composites are classified on the basis of matrix, reinforcement, and morphology. Polymers, ceramics, and metals are examples of matrices, whereas fibers and particulates are examples of reinforcements. Here, we discuss polymer composites used in EMI applications. The most common polymer matrices used are polyolefin, polyester, vinyl, ester, epoxy, phenolic, polyimide, polyamide, polypropylene, and poly ether ether ketone (PEEK) [24]. Recently developed polymer composites show outstanding performance in EMI shielding and are used in many practical applications, such as aerospace, military equipment, electronics, and medical equipment. Different types of polymer composites used include metal-coated polymers, conducting polymers, carbon nanotube–based composites, and graphene-based composites [25–27].

8.4.2.1 Metal-Coated Polymers

A metal coating on polymer fibers results in the metal-coated polymer shielding materials that have proven to be efficient in several applications, such as the automotive, aerospace, and computer industries. Different techniques are available to coat metals onto fibers, which include electroless plating, plasma treatment, foil laminates and tapes, ion plating, vacuum metallization, flame spraying, arc spraying, cathode sputtering, conductive paints and lacquers, and electroplating [28–32]. Metal-coated polymers possess several desirable properties, such as design flexibility, good formability, mechanical properties, coherent metal deposition, and excellent conductivity. The shielding ability of such composites depends on the interfacial adhesion between metal and polymer. Naturally, this adhesion is low, so several methods are used to increase interfacial adhesion, which include chemical modifications and plasma modifications. Apart from these treatments, homogeneous metal

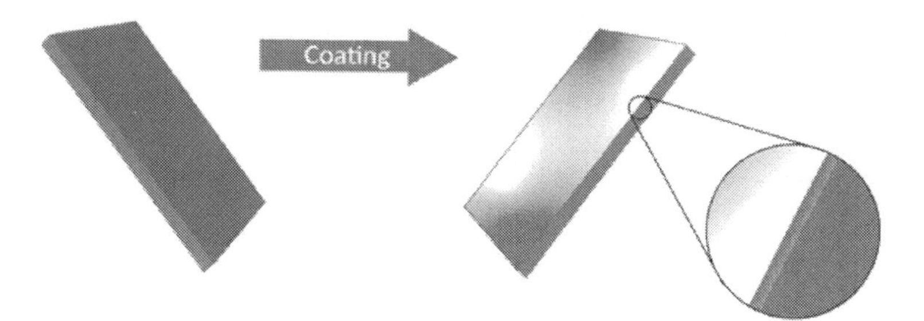

Figure 8.3 Schematic illustration of metal-coated polymer fibers.

deposition and the conductivity of metal also affect the performance of the composites.

8.4.2.2 Conducting Polymer–Based Materials

The metallike properties of molecularly doped polyacetylene were first proposed in 1977 [1]. Thus, conducting polymers, a new class of materials with exceptional properties such as the reflection and absorption of electromagnetic radiation over a wide range, the dissipation of electric charge, and applications in energy storage and display devices, were introduced [33–35].

Nowadays, conductive polymers are used in textiles to protect humans from electromagnetic radiation. These materials can be used as alternatives to metals in EMI shielding as they provide EMI shielding from both the absorption and reflection of waves, whereas metals only reflect them. Intrinsically conducting polymers such as polypyrrole- (PPy) and polyaniline-coated fabrics with natural or synthetic fibers are commonly used [36,37].

8.4.2.3 Nanofiller-Based Composites

As mentioned previously, reinforcing materials used in composites play a major role in the performance of the composites. Nanofillers are a class of nano-sized filler materials with distinct properties such as high surface area and novel electrical, thermal, dielectric, magnetic, and/or mechanical properties [38,39]. Nanofiller-reinforced composites show better shielding performance compared with other filler materials. Nanofillers such as carbon nanotubes (CNT) and graphene-based composites have exhibited good shielding capabilities. These fillers can be incorporated into a wide variety of polymers, including thermoplastics and thermosets. These composites are capable of absorbing and reflecting the electromagnetic radiation, which results in improved shielding performance. CNT is an allotrope of carbon and can be described as rolled graphene sheets in a pattern or as multiple concentric cylinders. CNTs show exceptional conductivity, high strength, good thermal properties, and good percolation in polymer matrices. CNT-based nanocomposite shielding materials find applications in areas such as aerospace, defense, and electronic devices.

Another major class of nanomaterials used is graphene. Two-dimensional single- or multilayered graphene sheets are used. A graphene sheet consists of sp^2-bonded carbon atoms. Graphene sheets also show exceptional conductivity, thermal properties, and mechanical properties and are cheaper than CNTs. Further, the performance of graphene-based shielding materials depends on interfacial adhesion between the graphene sheets and the polymer. Therefore, the ends of graphene sheets are factionalized to achieve good interfacial properties. But sometimes this results in decreased conductivity of the graphene sheets.

8.5 CONCLUSION

In the age of rapid technological development, electronic and communication devices play a very important role. These systems also form the basis of airplane and spacecraft communication systems. This chapter deals with the EMI shielding materials used to effectively counter the radiation occurring from both the earth and space. EMI is a critical issue that arises from household electronics, telecommunication systems, and natural phenomena like lightening, the sun, and the universe. This radiation interferes with electronic systems and can even lead to disasters if not properly countered. But scientists have developed efficient EMI shielding materials through active research and development. It is of great importance to understand the radiation environment and select appropriate materials for different applications. Metal-based enclosures, metal-coated polymer composites, intrinsically conducting polymers, nanofiller-based composites, and conducting filler–based composites are all identified as excellent materials for EMI shielding applications.

REFERENCES

1. Mishra, R.K., Thomas, M.G., Abraham, J., Joseph, K. and Thomas, S. (2018). Electromagnetic interference shielding materials for aerospace application: A state of the art. *Advanced Materials for Electromagnetic Shielding: Fundamentals, Properties, and Applications*, 327–365.

2. Messenger, G.C. and Ash, M.S. (1986). *The Effects of Radiation on Electronic Systems*. New York: Van Nostrand Reinhold.
3. Ma, T.P. and Dressender, P.V. (eds.) (1989). *Ionizing Radiation Effects in MOS Devices and Circuits*. New York: Wiley.
4. Corliss, W.R. (1968). Space radiation. United States Atomic Energy Commission Office of Information Services.
5. Bhat, B.R. and Sahu, R.P. (1993). Radiation shielding of electronic components in INSAT-2. *Journal of Spacecraft Technology* 3: 36.
6. Jones, S.A. (1997). U.S. Patent No. 5,670,742. Washington, DC: U.S. Patent and Trademark Office.
7. Kroll, M.W. (2003). U.S. Patent No. 6,580,915. Washington, DC: U.S. Patent and Trademark Office.
8. Kovacevic, I.F., Friedli, T., Muesing, A.M. and Kolar, J.W. (2014). 3-D electromagnetic modeling of EMI input filters. *IEEE Transactions on Industrial Electronics* 61 (1): 231–242.
9. Golio, M. (ed.) (2000). *The RF and Microwave Handbook*. Boca Raton, FL: CRC press.
10. Shooman, M.A. (1994). Study of occurrence rates of electromagnetic interference (EMI) to aircraft with a focus on HIRF (external) high intensity radiated fields. NASA Contractor Report 194895.
11. Liong, S. (2005). A multifunctional approach to development, fabrication, and characterization of Fe3O4 composites. Dissertation, Georgia Institute of Technology.
12. Mark, H.F., Bikales, N., Overberger, C.G., Menges, G. and Kroschwitz, J.I. (1987). *Encyclopedia of Polymer Science and Engineering*, Vol. 10. Hoboken, NJ: Nonwoven Fabrics to Photopolymerization.
13. Rothon, R.N. and Hancock, M. (1995). General principles guiding selection and use of particulate materials. In: *Particulate-Filled Polymer Composites*, 1–42. England: Longman Scientific & Technical.
14. Thostenson, E.T. and Chou, T.W. (2003). On the elastic properties of carbon nanotube-based composites: Modelling and characterization. *Journal of Physics D: Applied Physics* 36 (5): 573.
15. Hale, J. (2006). Boeing 787 from the ground up. *Aero* 4: 17–24.
16. Zhang, R.X., Ni, Q.Q., Natsuki, T. and Iwamoto, M. (2007). Mechanical properties of composites filled with SMA particles and short fibers. *Composite Structures* 79 (1): 90–96.
17. Rafiee, M.A., Rafiee, J., Wang, Z. et al. (2009). Enhanced mechanical properties of nanocomposites at low graphene content. *ACS Nano* 3 (12): 3884–3890.
18. Karayacoubian, P., Yovanovich, M.M. and Culham, J.R. (2006, March). Thermal resistance- based bounds for the effective conductivity of composite thermal interface materials. In Semiconductor Thermal Measurement and Management Symposium, 2006 IEEE Twenty- Second Annual IEEE, IEEE, pp. 28–36.

19. Yu, P., Chang, C.H., Su, M.S. et al. (2010). Embedded indium-tin-oxide nanoelectrodes for efficiency and lifetime enhancement of polymer-based solar cells. *Applied Physics Letters* 96 (15): 153307.
20. Gopakumar, D.A., Pai, A.R., Pottathara, Y.B., Pasquini, D., Carlos de Morais, L., Luke, M., . . . Thomas, S. (2018). Cellulose nanofiber-based polyaniline flexible papers as sustainable microwave absorbers in the X-band. *ACS Applied Materials & Interfaces* 10 (23): 20032–20043.
21. Gopakumar, D.A., Pai, A.R., Pottathara, Y.B., Pasquini, D., de Morais, L.C., Khalil, H.P.S.A., . . . Thomas, S. (2021). Flexible papers derived from polypyrrole deposited cellulose nanofibers for enhanced electromagnetic interference shielding in gigahertz frequencies. *Journal of Applied Polymer Science* 138 (16): 50262.
22. Pai, A.R., Paoloni, C. and Thomas, S. (2021). Nanocellulose-based sustainable microwave absorbers to stifle electromagnetic pollution. In: *Nanocellulose Based Composites for Electronics*, 237–258. Amsterdam, Netherlands: Elsevier.
23. Chung, W. Radiation - Atomic Radiation (1995–2018) available at: www.projectrho.com/ public_html/rocket/radiation.php 20 COSMIC, RAYS. (2005) "PDG Revi.
24. Bhat, B.R. and Sahu, R.P. (1993). Radiation shielding of electronic components in INSAT2. *Journal of Spacecraft Technology* 3: 36.
25. Bigg, D.M. and Bradbury, E.J. (1981). Conductive polymeric composites from short conductive fibers. In: *Conductive Polymers* (ed. R.B. Seymour), 23–38. New York: Springer.
26. Rai, M. and Yadav, R.K. (2014). Characterization of shielding effectiveness of general metallized structure. *International Journal of Wireless and Microwave Technologies (IJWMT)* 4 (5): 32.
27. Trostyanskaya, E.B. (1995). *Polymer Matrix Composites, Soviet Advanced Composites Technology Series* (ed. R.E. Shalin), 1–91. London: Chapman & Hall.
28. Manenq, F., Carlotti, S. and Mas, A. (1999). Some plasma treatment of PET fibres and adhesion testing to rubber. *Die Angewandte Makromolekulare Chemie* 271 (1): 11–17.
29. Inagaki, N., Tasaka, S., Narushima, K. and Mochizuki, K. (1999). Surface modification of tetrafluoroethylene-perfluoroalkyl vinyl ether copolymer (PFA) by remote hydrogen plasma and surface metallization with electroless plating of copper metal. *Macromolecules* 32 (25): 8566–8571.
30. Sun, R.D., Tryk, D.A., Hashimoto, K. and Fujishima, A. (1998). Formation of catalytic Pd on ZnO thin films for electroless metal deposition. *Journal of the Electrochemical Society* (10): 3378–3382.
31. Chu, S.Z., Sakairi, M., Takahashi, H., and Qui, Z.X. (1999). Local deposition of Ni-P alloy on aluminum by laser irradiation and electroless plating. *Journal of the Electrochemical Society* 146 (2): 537–546.

32. Karmalkar, S. and Banerjee, J. (1999). A study of immersion processes of activating polished crystalline silicon for autocatalytic electroless deposition of palladium and other metals. *Journal of the Electrochemical Society* 146 (2): 580–584.
33. Kaynak, A., Unsworth, J., Beard, G. and Clout, R. (1993). Study of conducting polypyrrole films in the microwave region. *Materials Research Bulletin* 28 (11): 1109–1125.
34. Kaynak, A. (1996). Electromagnetic shielding effectiveness of galvanostatically synthesized conducting polypyrrole films in the 300–2000 MHz frequency range. *Materials Research Bulletin* 31 (7): 845–860.
35. Kaynak, A., Unsworth, J., Clout, R. et al. (1994). A study of microwave transmission, reflection, absorption, and shielding effectiveness of conducting polypyrrole films. *Journal of Applied Polymer Science* 54 (3): 269–278.
36. Florio, L. and Amelia Carolina Sparavigna (2004). Polypyrrole/PET textiles as materials for safety through electromagnetic shielding. INF Meeting, Genova, Abstracts p. 193.
37. Colaneri, N.F. and Shacklette, L.W. (1992). EMI shielding measurements of conductive polymer blends. *IEEE Transactions on Instrumentation and Measurement* 41 (2): 291–297.
38. Chung, D.D.L. (2001). Electromagnetic interference shielding effectiveness of carbon materials. *Carbon* 39 (2): 279–285.
39. Choudary, V. and Dhawan, S.K. (2012). Polymer based nanocomposites for electromagnetic interference (EMI) shielding. In: *EMI Shielding Theory and Development of New Materials* (eds. M. Jaroszewski and J. Ziaja), 67–100. India: Research Signpost.

9

OVERVIEW OF NANOSTRUCTURED MATERIALS FOR ELECTROMAGNETIC INTERFERENCE SHIELDING

P. A. Nizam, T. Binumol, and Sabu Thomas

⇒ **CONTENTS**

9.1 Introduction 99
9.2 Intrinsically Conducting Polymers (ICPs) 100
9.3 Carbon-Based Fillers 102
9.4 Polymer/Inorganic Nanocomposites 105
9.5 Conclusion 106
References 107

9.1 INTRODUCTION

Nanostructured materials are an essential ingredient for future recipes for electromagnetic interference shielding materials. Their prominence in future applications is conspicuous, which can be attributed to their excellent properties such as tunable physiochemical characteristics, wettability, thermal and electrical conductivity, light absorption, melting point, and catalytic activity. A plethora of research is already being carried out in various applications like water purification[1], edible coatings[2], drug delivery, EMI shielding[3], supercapacitors[4], and fuel cells[5], employing wide classes of nanoparticles. These particles are categorized as nano when one of their dimensions is in the nano scale(1–100 nm). The British standard institution defines nanomaterials as materials with any physical dimension(s) in nano scale or internal structure in the nano scale dimension.

Five categories of nanomaterials have been identified so far: (1) carbon-based, (2) inorganic-based, (3) organic-based, (4) composite-based, and (5) bio-based[6] materials. Nanomaterials for EMI shielding

DOI: 10.1201/9781003217312-9

primarily employ carbon-based and inorganic materials as fillers, as they have proved to be excellent contributors to these applications. The most common carbon-based fillers include carbon nanotubes, graphenes, and carbon black, and the inorganic classes mostly include metal and metal oxide nanoparticles. Other fillers, such as intrinsically conducting polymers, polyaniline, and polypyrrole, offer transparent EMI shielding as they are electrically tunable, cost-efficient, and mechanically stable. Apart from fillers, nanopolymers are used as matrices for shielding applications; for example, nanocellulose has good hierarchical structure and is used as a matrix for EMI shielding applications.

Inorganic nanoparticles and carbon-based materials are dominant in shielding applications, as they have high conductivity and mobile electrons for interaction with the electric field of the radiation. Polymers are less effective when used in their native form, unless they are conducting. Different provisions are employed to process polymers in order to utilize them for different applications. The presence of an electrical and a magnetic constituent enhances the absorption contribution of the shielding due to interactions between the material and the electrical or magnetic field of the radiation. This chapter discusses various nanostructured materials employed for EMI shielding applications and their pros and cons.

9.2 INTRINSICALLY CONDUCTING POLYMERS (ICPS)

Intrinsically conducting polymers have proved to be excellent materials for shielding applications. Delocalization of p-electrons in the conjugated structure of an ICP has facilitated its importance as it provides a unique electronic property that can be tuned by doping or de-doping. The conducting nature of the ICPs can be attributed to the presence of mobile charges such as polarons, bipolarons, and dipoles present in their backbones. However, extensive delocalization of p-electrons hinders their processability[6]. Often the inferior properties of these materials, such as mechanical strength and electrical properties, are enhanced by the incorporation of fillers such as metals, metal oxides, and carbon nanomaterials. Conducting polymers are conjugated polymers that, upon doping, exhibit electrical conductivity. This doping process with either electron acceptors or donors creates a band structure to escalate their conductivity by many orders. Some of the common examples of ICPs include polyacetylene, polypyrrole, polyaniline, poly(*p*-phenylene vinylene), and polythiophene[7].

The strong tailorability of their electrical conductivity by modulating their doping level, oxidation state, chemical structure, and morphology makes them superior candidates for EMI shielding applications. Undoped polymers can act as an insulator or as a semiconductor because their conductivity lies in the range from 10^{-10} to 10^{-5} S/cm. Nonetheless, controlled doping magnifies their conductivity to the semiconductor or metallic conductivity range from 10^{-5} to 10^{5} S/cm. The intrinsic conductivity of 100 MHz to 20 GHz in the discipline of microwaves of these conjugated polymers promotes them as viable materials for shielding. Tunable conductivity (between insulating and metallic limits), adjustable permittivity and permeability via synthetic means, low density, noncorrosiveness, nominal cost, simple processing, and controllable electromagnetic properties strengthen them as futuristic shielding materials for a variety of technocommercial applications[8].

The properties of a large number of available shielding materials studied reveal that no single-phase material can cover all of the aspects of the shield (e.g. coefficient of absorption, volume, thickness, broadband response) to provide the desired performance level under different environments and applications. These drawbacks are overcome by processing ICPs with other nanomaterials, where the ICP is being employed as a filler as well as a matrix. For example, the nanocomposite of polyaniline and CNF, where PANI is the filler, is fabricated in the form of an aerogel as well as a film. The lightweight hybrid aerogel exhibits better shielding efficiency due to its large surface area when compared to films. Although PANI is employed in both studies, it exhibits different efficiency depending on the matrix.

ICP as a matrix has the advantages of design flexibility, interaction with fillers, and microwave nontransparency[9]. The filler, such as conducting, magnetic, or dielectric nanoparticles, can be incorporated into ICPs via an *in situ* or *ex situ* process. The latter has the disadvantage of poor dispersion of nanoparticles, which leads to agglomeration and further results in inferior electrical and electromagnetic properties. In contrast, the electrical characteristics may be rigorously controlled by using the *in situ* incorporation strategy, which involves performing polymerization under controlled conditions and in the presence of specific dopants and fillers.

Polyaniline has the advantages of wider configurations, high environmental stability, adjustable electrical and optical properties, easy conductivity modulation by altering oxidation and protonation states, and specific doping methods[10]. PANI is studied for its shielding efficacy in the MHz and GHz frequency range in various forms, which include thin films, multilayer films, textiles, adhesives, nanoparticles, and composites. These materials exhibit an EMI SE in the range of 16–50 dB at 0.1 GHz frequency[11]. Polypyrrole is another promising conducting polymer with intriguing electrical characteristics[12]. It is both thermally and chemically stable, as well as easy to synthesize. It suffers from poor processability and mechanical strength, however, along with infusibility and insolubility. Although polypyrrole polymers are poor conductors, when properly oxidized they become electrically conducting and their properties strongly depend on the preparation procedure. They exhibit a conductivity of 10^{-3} S/cm when properly doped.

Polyacetylene, synthesized in 1974, was the first material used in research studies after the development of doping. Polyacetylene was exposed to dopant chemicals, oxidizing or reducing agents, electron donors, or electron receptors to generate conductive polyacetylene polymers. Although the discovery of polyacetylene sparked the creation of conductive polymers, it has no significant commercial uses. ICPs have contributed to significant advancements in the field of EMI shielding applications[13].

9.3 CARBON-BASED FILLERS

Carbon materials (e.g. coke, graphite, graphene, carbon fiber, carbon nanofiber, and carbon nanotube) are electrically conductive as well as good absorbers of electromagnetic radiation over a wide frequency range. Carbon-based fillers include carbon black (CB), carbon nanofibers (CNFs), carbon nanotubes (CNTs), and graphene. Among these fillers, CB is considered to be a primitive material used for EMI shielding applications. Carbon blacks is obtained by the partial combustion or thermal decomposition of hydrocarbons. CB essentially consists of elemental carbon in the form of particles fused into aggregates. They are normally considered to be semiconductors due to their graphitic nature. Their agglomeration characteristics and poor dispersibility limit their application in EMI shielding[14].

Carbon nanofibers (CNFs) are cylindrical nanostructures containing graphene layers that are formed into cups, cones, or plates and have an average length of 0.5–200 nm and average diameter of 10–500 nm. Their low density, high aspect ratio, large specific surface area, good dimensional stability, modulus, flexibility, thermal stability, mechanical strength, and excellent thermal and electrical conductivities enable them to be employed in applications such as EMI shielding, composites for vehicles, and aerospace. These materials exhibit an EMI SE of −40 dB with a 14 vol% polymer composites within the K-band frequency range. Moreover, they also exhibit good mechanical properties and thermal stability, which aid their use in commercial applications. Their increased thermal stability can be attributed to the barrier effect of the thermal transfer of CNF well dispersed in the polymer matrix[15].

Carbon nanotubes (CNTs) are carbon allotropes having a cylindrical or tubular nanostructure made up of six-membered carbon rings. These cylindrical carbon molecules offer unique advantages in nanotechnology, electronics, optics, and other materials science and technology domains. Carbon nanotubes are also used as additives in a variety of structural materials due to their exceptional thermal conductivity and mechanical and electrical properties. There are two types of nanotubes: single-walled carbon nanotubes (SWCNTs)[16] and multiwalled carbon nanotubes (MWCNTs)[17]. Due to their smaller diameter and greater aspect ratio, SWCNTs have different electrical characteristics than MWCNTs. As a result, when incorporates into polymer composites, they display differing EMI SE properties. The primary disadvantages of SWCNTs include the complex production process, extremely high conductivity, and poor magnetic properties. These characteristics restrict the use of SWCNTs as good microwave absorbers. MWCNTs are the carbon nanotubes used most often. MWCNT is made up of many layers of graphite that have been overlaid and rolled into a tubular form. The structural defects that emerge in MWCNTs during the manufacturing process are responsible for their unique optical, electrical, and other properties. In terms of tensile strength and elastic modulus, carbon nanotubes are the strongest and stiffest materials found to date. CNTs offer substantial benefits over conventional carbon-based fillers because of their low weight, smaller diameter, high percolation and aspect ratio outstanding conductivity, and superior mechanical strength. High EMI SE may be easily attained at relatively low concentrations in polymer matrices.

Graphene is the 2D Nobel Prize–winning material with a single layer of sp^2-hybridized carbon atoms. The percolation threshold can be attained at a very low concentration due to the atomically thick structure and ultrahigh specific surface area[18]. Graphene is a very promising candidate for electronics, sensors, optoelectronics, structural materials, energy storage, and EMI shielding applications due to its exceptional charge carrier mobility, excellent thermal and electrical conductivity, and unique mechanical, optical, and magnetic properties. Graphene/polymer composites have been at the forefront of functional shielding materials, showing considerable promise for overcoming issues related to physical and mechanical performance, durability, and usefulness in the realm of electromagnetic shielding. However, their extremely high carrier mobility and lack of surface functions can be deleterious to EMI absorption, resulting in impedance mismatching between the material and air. As a result, graphene derivatives such as graphene oxide (GO) and reduced graphene oxide (rGO) have been widely employed in practical applications as alternatives to graphene.

Graphenes are homogeneously mixed in polymer matrices and exhibit an electrical conductivity of 10^2 S/m and high EMI shielding. But they face the disadvantage of heating due to high electrical conductivity. Graphene oxide is the oxidized form of graphene with added functional groups such as epoxy, carbonyl, carboxyl, ketone, or diol on the edges and basal planes. Graphene oxide's exceptional compatibility with polymers makes it an appealing filler for the manufacture of composites with greatly improved electrical and thermal conductivity, tensile strength, and elasticity. The oxygen-containing functional groups of GO contribute to the outstanding mechanical and reinforcing characteristics of polymer composites by increasing interfacial bonding and transferring stresses from the polymer matrix to the filler and vice versa[19]. Reduced graphene oxide (rGO) is a heterogeneous structure composed of a graphene-like basal plane embellished with extra structural flaws and filled with oxidized chemical groups and heteroatoms. rGO's graphene-like characteristics make it an attractive material for use in sensors and environmental, biological, or catalytic applications, as well as in optoelectronics, storage devices, and composite materials. Furthermore, the rGO sheet's functional groups and structural defects contribute to improved impedance mismatch, defect polarization relaxation, and electronic dipole polarization. These materials exhibit shielding effectiveness of 25–55 dB, depending on the matrix and the inorganic fillers added[20].

Other forms of graphene-like materials such as graphene nanoplatelets and graphene nanoribbons have been widely investigated. Graphene nanoplatelets (GNPs) are one-of-a-kind nanoparticles made up of tiny stacks of 10–30 graphene sheets with a platelet shape, similar to carbon nanotubes but in a planar form. GNPs' unusual size and platelet form make them particularly useful for providing barrier qualities, while their pure graphitic composition gives them exceptional thermal and electrical properties[21]. The GNP 2D planar structure, which reduces phonon scattering at the polymer/nanofiller interface, is responsible for the remarkable improvement in heat conductivity of the GNP-based polymer composites. Their easy processability, economic viability, good dispersion, and other properties are superior to those of CNTs. Graphene nanoribbons (GNRs) are long, thin graphene strips of sp^2 carbon atoms having a width of 12–20 nm. GNRs exhibit remarkable charge transfer and electrical capabilities because of their numerous edge groups, ultrahigh aspect ratio, and larger effective surface area. GNRs have become particularly attractive nanofillers for a variety of applications, including energy storage, transistors, polymer nanocomposites, high-strength materials, and EMI shielding, due to their excellent mechanical, chemical, thermal, and magnetic characteristics[22]. Carbon-based fillers are the materials that have been explored the most as for EMI shielding applications and have a prosperous future due to their good conductivity, light weight, and flexibility.

9.4 POLYMER/INORGANIC NANOCOMPOSITES

Various criteria are employed for polymer composites, such as blending and doping, depending upon the end application and desired properties. As the current field of study has advanced at a breakneck pace, fresh and modern views about new developments have emerged. Multicomponent structures have been employed in many applications with all of the required properties. Multicomponent structures are hybrid materials that incorporate several elements that contribute to electromagnetic shielding efficiency into a single structural unit, offering the highest level of EMI shielding polymer composites. A carbon-based filler along with inorganic nanoparticles to enhance the properties of the composite is an example of a multicomponent composite that has the advantages of both of the fillers[23].

Inorganic metals and metal oxide nanoparticles are promising nanofillers for EMI shielding applications. They enhance the SE when used in conjunction with other hybrid composites. Carbon-based fillers along with metals, ferrites, and metal oxides escalate the tunability of physicomechanical properties, as well as the permittivity, permeability, thickness, and electrical conductivity.

Variations of inorganic metal fillers with carbon-based fillers may be a potential approach to coupling the magnetic properties of magnetic metals with the dielectric properties of carbon inclusions to change EM wave attenuation. Carbonyl iron particles along with CB incorporated into the polymer matrix improves magnetic characteristics and shielding performance (20–27 dB). The improved EMI SE in these types of hybrid fillers is due to the percolation network, via the incorporation of the fillers, and also due to the formation of magnetic and electrical diploes. Liang et al. used a sol-gel template to create a 3D reduced graphene oxide/silver platelet foam (rGO/AgPs) with many spherical hollow structures. By combining this foam with epoxy resin, 3D EP/rGO/AgPs nanocomposites with strongly regular, segregated structures were created. The 3D EP/RGO/AgPs composites comprising 0.44 vol% rGO and 0.94 vol% AgPs demonstrate a maximum EMI SE of 58 dB in the X-band frequency range due to the interconnected, hollow conducting networks of rGO/AgPs and the interfacial synergy between EP and rGO/AgPs (99.99% shielding of EM waves). When compared to the 3D EP/rGO, this is a 274% improvement[24]. Polyaniline incorporated with graphene and silver, decorated with nickel, composite was fabricated to understand the shielding effectives of metal decorated graphene. These composites exhibit a conductivity of 20.32 S/cm and SE of 30 dB. The uniform dispersions escalate the pathways for conduction, and the presence of metal nanoparticles improves their conductivity[25]. Studies have indicated that metal nanoparticles incorporated in carbon-based fillers are excellent materials for EMI shielding.

9.5 CONCLUSION

The current chapter provided an overview of common fillers used in the applications of EMI shielding. Intrinsically conducting polymers have been widely explored as a filler and as a matrix. These materials exhibit excellent stability, conductivity, and good EMI shielding efficiency. Their drawbacks in some properties can be addressed by the incorporation of carbon-based fillers, which includes a vast family such as carbon black, CNTs, and graphenes. Furthermore, these fillers' properties can be tuned by decoration with inorganic metal, metal oxide, and ferrite nanoparticles. Multicomponent systems that incorporate more than one filler are also a promising technology for EMI shielding applications.

REFERENCES

1. Nizam, P. A.; Arumughan, V.; Baby, A.; Sunil, M. A.; Pasquini, D.; Nzihou, A.; Thomas, S.; Gopakumar, D. A. Mechanically Robust Antibacterial Nanopapers Through Mixed Dimensional Assembly for Anionic Dye Removal. *J. Polym. Environ.* **2020**, *28* (4), 1279–1291.
2. Rose Joseph, M.; Nizam, P. A.; Gopakumar, S.; Maria, H. J.; Vishnu, R.; Kalarikkal, N.; Thomas, S.; Vidyasagarana, K.; Anoop, E.V. Development and Characterization of Cellulose Nanofibre Reinforced Acacia Nilotica Gum Nanocomposite. *Ind. Crops Prod.* **2021**, *161* (December 2020), 113180.
3. Gopakumar, D. A.; Pai, A. R.; Pottathara, Y. B.; Pasquini, D.; Carlos De Morais, L.; Luke, M.; Kalarikkal, N.; Grohens, Y.; Thomas, S. Cellulose Nanofiber-Based Polyaniline Flexible Papers as Sustainable Microwave Absorbers in the X-Band. *ACS Appl. Mater. Interfaces.* **2018**, *10* (23), 20032–20043.
4. Rajeevan, S.; John, S.; George, S. C. Polyvinylidene Fluoride : A Multifunctional Polymer in Supercapacitor Applications. *J. Power Sources.* **2021**, *504* (January), 230037.
5. Nizam, P. A., et al. Nanocellulose-based composites: fundamentals and applications in electronics. *Nanocellulose Based Composites for Electronics.* Elsevier, **2021**. 15–29.
6. Sathyanarayana, D. N.; Chemistry, P.; Division, M. T. Pergamon Conducting Polyaniline Blends and Composites. *Prog. Polym. Sci.* **1998**, *23* (97), 993–1018.
7. Jiang, D.; Murugadoss, V.; Wang, Y.; Lin, J.; Ding, T.; Wang, Z.; Shao, Q.; Wang, C.; Liu, H.; Lu, N.; Wei, R.; Subramania, A.; Guo, Z. Electromagnetic Interference Shielding Polymers and Nanocomposites—A Review. *Polym. Rev.* **2019**, *211*, 108827.
8. Kim, C.; Kim, M. Intrinsically Conducting Polymer (ICP) Coated Aramid Fiber Reinforced Composites for Broadband Radar Absorbing Structures (RAS). *Compos. Sci. Technol.* **2021**, *211*, 108827.
9. Rajeevan, S.; John, S.; George, S. C. The Effect of Poly(Vinylidene Fluoride) Binder on the Electrochemical Performance of Graphitic Electrodes. *J. Energy Storage.* **2021**, *39* (January), 102654.
10. Pai, A. R.; Binumol, T.; Gopakumar, D. A.; Pasquini, D.; Seantier, B.; Kalarikkal, N.; Thomas, S. Ultra-Fast Heat Dissipating Aerogels Derived from Polyaniline Anchored Cellulose Nanofibers as Sustainable Microwave Absorbers. *Carbohydr. Polym.* **2020**, *246*, 116663.
11. Dar, M. A.; Majid, K.; Farukh, M.; Dhawan, S. K.; Kotnala, R. K.; Shah, J. Electromagnetic Attributes a Dominant Factor for the Enhanced EMI Shielding of PANI/Li0.5Fe2.5−xGdxO4 Core Shell Structured Nanomaterial. *Arab. J. Chem.* **2019**, *12* (8), 5111–5119.

12. Gopakumar, D. A.; Pai, A. R.; Pottathara, Y. B.; Pasquini, D.; de Morais, L. C.; Khalil, H. P. S. A.; Nzihou, A.; Thomas, S. Flexible Papers Derived from Polypyrrole Deposited Cellulose Nanofibers for Enhanced Electromagnetic Interference Shielding in Gigahertz Frequencies. *J. Appl. Polym. Sci.* **2021**, *138* (16), 1–11.
13. Lee, C. Y.; Kim, H. M.; Park, J. W.; Gal, Y. S.; Jin, J. I.; Joo, J. AC Electrical Properties of Conjugated Polymers and Theoretical High-Frequency Behavior of Multilayer Films. *Synth. Met.* **2001**, *117* (1–3), 109–113.
14. Revati, R.; Abdul Majid, M. S.; Yahud, S.; Cheng, E. M.; Ridzuan, M. J. M. Electromagnetic Interference (EMI) Shielding Effectiveness (SE) of Unsaturated Polyester Resin with Carbon Black Fillers. *J. Telecommun. Electron. Comput. Eng.* **2018**, *10* (1–17), 119–124.
15. Kang, K.; Kim, H.; Choi, E.; Shim, H. J.; Cui, Y.; Gao, Y.; Huh, H.; Pyo, S. G. Electro-Magnetic Interference Shielding Effect of Electrospun Carbon Nanofiber Web Containing BaTiO3 and Fe3O4 Nanoparticles. *J. Nanosci. Nanotechnol.* **2017**, *17* (10), 7689–7694(6).
16. Joseph, N.; Janardhanan, C.; Sebastian, M. T. Electromagnetic Interference Shielding Properties of Butyl Rubber-Single Walled Carbon Nanotube Composites. *Compos. Sci. Technol.* **2014**, *101*, 139–144.
17. Gahlout, P.; Choudhary, V. EMI Shielding Response of Polypyrrole-MWCNT/Polyurethane Composites. *Synth. Met.* **2020**, *266*, 116414.
18. Raagulan, K.; Kim, B. M.; Chai, K. Y. Recent Advancement of Electromagnetic Interference (EMI) Shielding of Two Dimensional (2D) MXene and Graphene Aerogel Composites. *Nanomaterials.* **2020**, *10* (4), 702.
19. Lin, S.; Ju, S.; Zhang, J.; Shi, G.; He, Y.; Jiang, D. Ultrathin Flexible Graphene Films with High Thermal Conductivity and Excellent EMI Shielding Performance Using Large-Sized Graphene Oxide Flakes. *RSC Adv.* **2019**, *9* (3), 1419–1427..
20. Chen, Y.; Pötschke, P.; Pionteck, J.; Voit, B.; Qi, H. Multifunctional Cellulose/RGO/Fe3O4 Composite Aerogels for Electromagnetic Interference Shielding. *ACS Appl. Mater. Interfaces* **2020**, *12* (19), 22088–22098.
21. Bhaskaran, K.; Bheema, R. K.; Etika, K. C. The Influence of Fe3O4@ GNP Hybrids on Enhancing the EMI Shielding Effectiveness of Epoxy Composites in the X-Band. *Synth. Met.* **2020**, *265*, 116374.
22. Arjmand, M.; Sadeghi, S.; Navas, I. O.; Keteklahijani, Y. Z.; Dordanihaghighi, S.; Sundararaj, U. Carbon Nanotube

versus Graphene Nanoribbon: Impact of Nanofiller Geometry on Electromagnetic Interference Shielding of Polyvinylidene Fluoride Nanocomposites. *Polymers (Basel).* **2019**, *11* (6), 1064.

23. Sankaran, S.; Deshmukh, K.; Ahamed, M. B.; Khadheer Pasha, S. K. Recent Advances in Electromagnetic Interference Shielding Properties of Metal and Carbon Filler Reinforced Flexible Polymer Composites: A Review. *Composites Part A: Applied Science and Manufacturing.* **2018**, *114*, 49–71.
24. Liang, C.; Song, P.; Qiu, H.; Zhang, Y.; Ma, X.; Qi, F.; Gu, H.; Kong, J.; Cao, D.; Gu, J. Constructing Interconnected Spherical Hollow Conductive Networks in Silver Platelets/Reduced Graphene Oxide Foam/Epoxy Nanocomposites for Superior Electromagnetic Interference Shielding Effectiveness. *Nanoscale.* **2019**, *11* (46), 22590–22598.
25. Chen, Y.; Li, Y.; Yip, M.; Tai, N. Electromagnetic Interference Shielding Efficiency of Polyaniline Composites Filled with Graphene Decorated with Metallic Nanoparticles. *Compos. Sci. Technol.* **2013**, *80*, 80–86.

INDEX

A

aerogels, 13, 27, 29, 37, 73

B

biodegradable polymer, 38
blow molding, 38
buckypaper (BP), 68

C

carbon black (CB), 10, 70, 78, 102
carbon fiber, 4
carbon nanofiber (CNFs), 60, 102, 103
carbonyl iron, 106
cement, 6
conducting polymer, 25, 68, 94, 100
conductivity, 4
cellulose-derived carbon aerogels (CCAs), 13
cellulose nanofiber (CNF), 30, 80
ceramics, 4, 6, 78, 82
chemical vapour deposition, 73
coke, 4, 102
copper nanowire, 60
core-shell structure, 17

E

electrostatic discharge (ESD), 1, 78, 88
epoxy, 93, 106

F

foam, 12
freeze drying, 29, 30

G

graphene 103, 104
  nanoplatelets, 104
  nanoribbons, 105
  oxide, 10
graphite, 5, 102

H

honeycomb structure, 18
hybrid composites, 6

L

lignin, 18

M

metal organic frameworks (MOFs), 78
multiwalled carbon nanotube, 11, 46, 103
MXenes, 6, 26, 27, 69, 78

P

percolation threshold, 4
permeability, 48, 61, 68, 105

permitivitty, 13, 48, 57, 105
polycaprolactone, 44
polycarbosilane, 79
polylactic acid, 40, 61

R

red mud, 81
reduced graphene oxide (rGO), 10, 48, 71, 106

S

sandwich structure, 15
scanning electron microscope, 29
segregated structure, 16
shape memory, 81
shielding effectiveness
  by absorption, 2, 37, 68
  by multiple internal reflection, 2, 37
  by reflection, 2, 37
skin effect, 3
specific shielding effectiveness, 13, 30
synthetic metals, 56

T

thermally reduced graphene (TRG) 80
thermoplastic polyurethane (TPU), 59

V

vapor phase polymerization (VPP), 17